Urban Freight Analytics

Urban Freight Analytics examines the key concepts associated with the development and application of decision support tools for evaluating and implementing city logistics solutions. New analytical methods are required for effectively planning and operating emerging technologies including the Internet of Things (IoT), Information and Communication Technologies (ICT), and Intelligent Transport Systems (ITS).

The book provides a comprehensive study of modelling and evaluation approaches to urban freight transport. It includes case studies from Japan, the US, Europe, and Australia that illustrate the experiences of cities that have already implemented city logistics, including analytical methods that address the complex issues associated with adopting advanced technologies such as autonomous vehicles and drones in urban freight transport.

Also considered are future directions in urban freight analytics, including hyperconnected city logistics based on the Physical Internet (PI), digital twins, gamification, and emerging technologies such as connected and autonomous vehicles in urban areas. An integrated modelling platform is described that considers multiple stakeholders or agents, including emerging organisations such as PI companies and entities such as crowd-shippers as well as traditional stakeholders such as shippers, receivers, carriers, administrators, and residents.

This book

- Presents procedures for evaluating city logistics technologies and policy measures
- Provides an overview of advanced modelling approaches, including agent-based models and machine learning
- Highlights the essential features of optimisation and simulation models applied to city logistics
- Discusses how models incorporating more uncertainty and dynamic data can be used to improve the sustainability and resilience of urban freight systems

The book is ideal for graduate students in civil and environmental engineering and logistics management, urban planners, transport engineers, and logistics specialists.

Urban Freight Analytics

Big Data, Models, and Artificial Intelligence

Eiichi Taniguchi
Russell G. Thompson
Ali G. Qureshi

CRC Press is an imprint of the
Taylor & Francis Group, an informa business

Front cover image: Eiichi Taniguchi, Russell G. Thompson, and Ali G. Qureshi

First edition published 2024
by CRC Press
2385 NW Executive Center Dr, Suite 320, Boca Raton, FL, 33431

and by CRC Press
4 Park Square, Milton Park, Abingdon, Oxon, OX14 4RN

CRC Press is an imprint of Taylor & Francis Group, LLC

Library of Congress Cataloging-in-Publication Data
Names: Taniguchi, Eiichi, author. | Thompson, Russell G., 1962- author. |
Qureshi, Ali G. (Ali Gul), author.
Title: Urban freight analytics : big data, models, and artificial
intelligence / Eiichi Taniguchi, Russell G. Thompson, and Ali G. Qureshi.
Description: First edition. | Boca Raton, FL : CRC Press, 2024. | Includes
bibliographical references and index.
Identifiers: LCCN 2023010167 | ISBN 9781032199368 (hardback) |
ISBN 9781032199375 (paperback) | ISBN 9781003261513 (ebook)
Subjects: LCSH: Urban transportation—Data processing—Planning. |
Urban transportation—Mathematical models. | Freight and freightage—
Planning—Mathematical models. | Big data. | Artificial intelligence—Data processing.
Classification: LCC TA1205 .T36 2024 | DDC 388.4—dc23/ENG/20230314
LC record available at https://lccn.loc.gov/2023010167

ISBN: 978-1-032-19936-8 (hbk)
ISBN: 978-1-032-19937-5 (pbk)
ISBN: 978-1-003-26151-3 (ebk)

DOI: 10.1201/9781003261513

Typeset in Sabon LT Std
by codeMantra

Contents

Preface

Urban freight transport is essential for sustainable and liveable cities. However, there are difficult problems relating to efficiency, negative environmental impacts, safety, and energy consumption associated with urban freight transport. City logistics has been proposed to overcome these complex problems and city logistics initiatives have been implemented in many cities around the world. Urban freight analytics allows us to model and understand the behaviour of stakeholders who are involved in city logistics and find innovative solutions based on emerging technologies including ITS (Intelligent Transport Systems), ICT (Information and Communication Technology), IoT (Internet of Things), big data and AI (Artificial Intelligence), autonomous vehicles, and robots.

This book contains recent advances in modelling and evaluating city logistics schemes. The aim of this book is to provide comprehensive knowledge and experience on city logistics to researchers, graduate students, and practitioners. Part 1 (Chapters 1–7) focuses on the methods of urban freight analytics, including the data collection using ICT, ITS and IoT and big data analyses, GIS (Geographical Information System), optimisation and multi-agent modelling with machine learning, reliability and resilience, and evaluation of policy measures. It highlights mathematical models that replicate the behaviour of multiple stakeholders and evaluate the city logistics policy measures. Part 2 (Chapters 8–12) addresses applications of urban freight analytics, including autonomous vehicles and robots, access management and pricing, environmental sustainability, and disruption of networks. It presents case studies from many cities around the world to help understand real problems and innovative solutions for urban freight transport. Furthermore, future perspectives focus on hyperconnected city logistics, emerging technologies such as digital twins, and integrated platforms for efficient and sustainable urban freight transport systems.

Chapter 1 presents basic concepts of city logistics, focusing on the use of emerging technologies such as ITS (Intelligent Transport Systems), ICT (Information and Communication Technology), IoT (Internet of Things), big data and AI (Artificial Intelligence), autonomous vehicles, and robots. Collaboration and coordination amongst stakeholders in city logistics are discussed. This chapter highlights an integrated platform based on the emerging technologies and collaboration between stakeholders for implementing city logistics policy measures. Concepts related to modelling city logistics are also addressed.

Chapter 2 provides methods for data collection and analysis for city logistics. It addresses data collection using IoT (Internet of Things), ICT (Information and Communication Technologies), and ITS (Intelligent Transport Systems), including VICS (Vehicle Identification and Communication Systems) and probe data. It discusses big data analyses in urban freight transport based on these data. Data sharing with the public sector and private companies is also discussed based on experiences in the Netherlands, Japan, France, Australia, the UK, and Sweden. Blockchain technology is discussed in terms of building trust for data sharing among stakeholders.

Chapter 3 demonstrates how spatial analysis tools can be employed to create information that can be used by decision makers to enhance the performance of urban freight systems. An overview of Global Position Systems (GPS) and Geographic Information Systems (GIS) is presented. GPS provides a cheap and automated means of logging the movement of freight vehicles allowing analysis of routes as well as their performance to be undertaken. GIS can be used to map a range of freight related data that can provide information that links various stakeholder interests.

Chapter 4 introduces a set of vehicle routing models that focus on different types of input data such as deterministic, dynamic, and stochastic. Several types of objective functions such as economic or environmental cost minimisation, and integration of facility location are also considered. These models are not only important from an optimisation point of view, but they also play a vital role in the evaluation of numerous city logistics schemes.

Chapter 5 presents multi-agent simulation methods with machine learning, which allows the behaviour and interaction of multiple stakeholders to be better understood. It encompasses reinforcement learning including Q-learning and adaptive dynamic programming (ADP) for supporting multi-agent simulation. Application of these models in urban delivery using urban consolidation centres (UCC) is discussed. This chapter also highlights decision support systems using multi-agent simulation including multi-actor multi-criteria analysis (MAMCA).

Chapter 6 presents reliability and resilience methods associated with urban freight transport. It highlights travel time reliability in stochastic vehicle routing and scheduling with time windows model which is often used for describing urban delivery with uncertainty. Incorporating risks of hazardous material transport in urban areas is addressed using the multi-objective optimisation models. Resilience in disasters is also described in the situation of the reduced capacity of road networks and logistics facilities in urban areas.

Chapter 7 presents a range of analytics methods that can be used for evaluating options to improve the sustainability of urban freight systems.

Financial analysis, multi-criteria assessment, and multi-objective optimisation problems are covered. Applications are described for a range of city logistics initiatives, including collaborative freight networks, electric freight vehicles, urban consolidation centres, and road tolls. This chapter describes techniques that can be used to highlight the trade-offs in the predicted performance of urban freight options and illustrate how sensitive these are to key assumptions within models. Procedures for illustrating the effects of options for multiple stakeholders with different objectives are presented.

Chapter 8 introduces the recent trend of incorporating un-manned vehicles in city logistics. Starting with a description of various types of autonomous delivery robots and drones, this chapter discusses their advantages and disadvantages. Critical issues associated with these advanced technologies such as their capacity and range, related regulations, and their acceptability by the public are discussed in this chapter. Examples from research and practice are included to understand the current state of integration of un-manned vehicles in city logistics theory and practice.

Chapter 9 describes how big data, analysis tools, and models can improve pricing and access management schemes for freight vehicles in urban areas. Descriptive analytics procedures presented include monitoring the usage of loading bays and loading docks. Prescriptive analytics methods including models developed for optimising the number of bays in loading docks and on-street as well as toll levels for trucks are described. This chapter highlights how analytical procedures can determine the best toll levels for urban freight vehicles as well as identifying optimal road pricing schemes.

Chapter 10 describes analytical methods for improving environmental sustainability in city logistics. It presents case studies on joint delivery systems using urban consolidation centres (UCC) in Tokyo, Japan. It focuses on the effectiveness of the pooling goods and the integration of UCC in joint delivery systems to reduce the distance travelled by trucks as well as the environmental impacts. It addresses cargo bikes and electric-assisted cargo bikes, which can reduce GHG (Green House Gas) and local emissions, but require micro depots or satellite facilities to transship goods from delivery trucks or vans to bikes.

Chapter 11 describes applications where streamlined logistics systems are adapted to the challenging conditions posed by large-scale disasters and pandemics. Concepts of vehicle routing and facility locations are adjusted by changing the focus from cost minimisation to coverage maximisation, delay minimisation, and ensuring equity in disaster relief distribution. The destruction caused by the disasters generates an enormous amount of debris. This chapter also presents models aimed at efficiently clearing and recovering road infrastructure.

Chapter 12 describes a range of models that will be required to facilitate the implementation of Hyperconnected City Logistics in the future. Agent-based modelling systems incorporating learning agents using AI are described. Optimisation procedures that incorporate more uncertainty, as well as interactions between multiple stakeholders, are outlined. Digital twin concepts and gamification methods are discussed. This chapter highlights the needs for enhancing optimisation models to incorporate more uncertainty and dynamic data.

This book is composed of 12 chapters, with each author making a major contribution to many chapters:

Eiichi Taniguchi: Chapters 1, 2, 5, 6, and 10
Russell G. Thompson: Chapters 3, 7, 9, and 12
Ali G. Qureshi: Chapters 2, 4, 8, and 11

We believe that this book will provide guidance for researchers and practitioners looking for insights relating to urban freight transport. The advanced modelling techniques and management methods that are presented in this book will allow innovative city logistics solutions to be developed for addressing the challenging and complicated urban freight transport issues. We hope that this book will promote further research and practice in an international and interdisciplinary manner in the area of city logistics.

Eiichi Taniguchi
Russell G. Thompson
Ali G. Qureshi

About the Authors

Professor Eiichi Taniguchi is a Professor Emeritus of Kyoto University, Japan. He has performed research on transport and logistics systems focusing on city logistics for sustainable, liveable, and resilient cities. He is the founder and President of the Institute for City Logistics. He published 23 books and over 240 academic papers on urban freight transport and Intelligent Transport Systems. His recent research covers city logistics based on big data, artificial intelligence, autonomous vehicles, robots, and the collaboration of stakeholders.

Professor Russell G. Thompson is a Professor in Transport Engineering at the University of Melbourne where he leads the Physical Internet Lab. Russell was a founding Director and has been the Vice President of the Institute for City Logistics based in Kyoto since 1999. His research areas are City Logistics, Physical Internet (PI), resilient transport systems, and Intelligent Transport Systems. Russell has published over 15 books and 150 refereed publications in the field of urban freight.

Associate Professor Ali G. Qureshi is an Associate Professor in the Department of Urban Management at Kyoto University, Japan. His research interests are related to the exact and heuristics optimisation of different variants of vehicle routing and facility location problems, their integration in different frameworks such as multi-agent systems, and their application in the evaluation of city logistics measures. He regularly contributes high-quality research papers in leading international and regional (Asia-Pacific) journals and in peer-reviewed conferences related with his areas of research.

Part I

Methods

Chapter 1

Introduction

1.1 CONCEPTS OF CITY LOGISTICS

Urban freight transport is an important element for the sustainable development of cities from the viewpoints of economy, environment, safety, security, energy, and health. However, there are many issues and challenges relating to urban freight transport, including traffic congestion, negative environmental impacts, crashes, energy consumption, and public health risk. The recent increase in e-commerce requires more efficient and reliable urban distribution of commodities. The problems of global greenhouse gas (GHG) emissions, local hazardous gas emissions, noise, and vibration create the need for more environmentally friendly urban freight transport. Safety, security, and public health in local communities should be taken into account in urban freight transport.

To tackle these difficult problems, the concept of city logistics has been proposed and implemented in many cities around the world. Taniguchi et al. (2001) defined city logistics as "the process for totally optimising the logistics and transport activities by private companies in urban areas while considering the traffic environment, the traffic congestion and energy consumption within the framework of a market economy".

City logistics provides innovative solutions for urban freight transport issues using emerging technologies such as Information and Communication Technologies (ICT), Intelligent Transport Systems (ITS), Internet of Things (IoT), big data, Artificial Intelligence (AI), autonomous vehicles, and robots (Taniguchi et al., 2020). Since many stakeholders such as shippers, freight carriers, administrators, and residents are involved in city logistics, collaboration amongst these stakeholders is essential for successful implementation of city logistics policy measures.

The research subject "city logistics" itself is interdisciplinary which requires a wide range of approaches from transport and logistics engineering, informatics, economics, public policy, geography, management, and others. The discipline of city logistics is aimed at balancing economic growth, social wellbeing, environmental friendliness, safety, security, and energy efficiency in urban areas. City logistics has three goals (Figure 1.1):

DOI: 10.1201/9781003261513-2

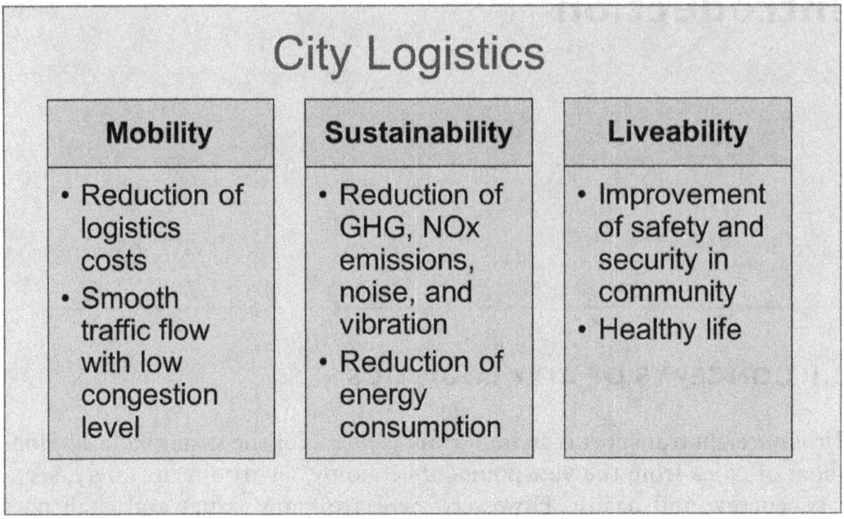

City Logistics

Mobility
- Reduction of logistics costs
- Smooth traffic flow with low congestion level

Sustainability
- Reduction of GHG, NOx emissions, noise, and vibration
- Reduction of energy consumption

Liveability
- Improvement of safety and security in community
- Healthy life

Figure 1.1 Three goals of city logistics.

mobility, sustainability, and liveability. The objectives of city logistics encompass the reduction of logistics costs as well as the decrease of traffic congestion, greenhouse gas (GHG) emissions and other negative environmental impacts, and energy consumption, and the improvement of safety and security in community. Therefore, interdisciplinary studies are required for tackling the complicated urban freight transport issues.

1.2 USE OF EMERGING TECHNOLOGIES

1.2.1 Overview

City logistics utilising recent innovative technologies including ITS (Intelligent Transport Systems), ICT (Information and Communication Technology), IoT (Internet of Things), big data and AI (Artificial Intelligence), autonomous vehicles, and robots are promising for evolving smart city logistics. Vehicle routing and scheduling with time windows models, multi-agent models, multi-actor multi-criteria models, and location routing models have been studied and applied in a large number of case studies for sustainable urban freight transport. Evaluating city logistics policy measures using these mathematical models is required to understand their effects before implementing them in advance. As several stakeholders including freight carriers, shippers, administrators, and residents have different objectives, evaluation using multiple criteria was often executed. A literature review is given below on the important topics in city logistics.

1.2.2 ICT, ITS, IoT, big data, and AI

Emerging technologies including ICT, ITS, IoT, big data, and AI are useful for establishing smart city logistics (Taniguchi & Thompson, 2014). These technologies allow an integrated platform for managing urban freight transport systems to be provided and operated in urban areas. The function of ITS and ICT includes:

a. Collecting data,
b. Storing data, and
c. Analysing data for improving existing urban freight transport systems.

We can collect precise data on the movements of freight vehicles and goods using ITS and ICT (see Chapter 2). For example, GPS (Global Positioning Systems) and other tele-communication technologies allow us to obtain location data of freight vehicles and goods every second. Such historical data which are stored long term in computers can be used for optimising routing and scheduling of delivery in urban areas. On-line data are also important for dynamic operation of freight vehicles considering traffic congestion.

Recently the IoT has become popular in production and logistics systems using sensors for measuring the location, acceleration, and weight of loads of freight vehicles, temperature within containers, as well as RFID (Radio Frequency Identification). IoT can connect a great number of production sites, distribution centres, freight vehicles, and commodities. Networks based on ITS, ICT, and IoT provide big data on urban freight transport systems.

Mehmood et al. (2017) analysed the effects of big data on city transport providing a new understanding of load sharing and optimisation in a smart city context. They demonstrated how big data can be used to improve transport efficiency and lower externalities in a smart city. Bibri (2018) highlighted IoT and related big data applications for smart sustainable cities, focusing on an analytical framework for data-centric applications enabled by IoT to advance environmental sustainability.

The important point of big data applications in city logistics is the collaborative operation of urban freight transport in sharing the capacity of vehicles, distribution centres, and information systems in the sharing economy. Big data related to urban freight transport are typically owned by private companies, but integration with public owned traffic and infrastructure data is needed for enhancing the operation of urban freight transport. Cleophas et al. (2019) discussed collaborative urban freight systems, identifying two key types of collaboration, vertical and horizontal. Vertical collaboration is referred to line-haul and last-mile or multi-echelon distribution, whereas in horizontal collaboration partners serve the same or at least over-lapping parts of the transport network. They considered strategic, tactical, and operational planning horizons as well as different degrees of information exchange and planning centralisation.

Taniguchi et al. (2018a) presented concepts of an integrated platform for innovative city logistics with urban consolidation centres (UCC) and transhipment points (TP). The concepts encompass joint delivery systems with shared use of pickup-delivery trucks, urban consolidation centres and transhipment points based on big data analytics. These systems involve public-private partnerships among shippers, freight carriers, UCC operators, municipalities, and regional planning organisations since the location of UCC and TP are highly related to urban land use plans and infrastructure provision as well as urban traffic management. ITS- and ICT-based communication systems for stakeholders are also essential for the efficient management of integrated platforms for both real-time and long-term operations. Integrated platforms can provide benefits such as improving the efficiency of urban deliveries and reducing CO_2 footprints (Dupas et al., 2020).

1.2.3 Autonomous vehicles and robots for last-mile delivery

Autonomous vehicles are promising technologies which can provide efficient delivery services for city dwellers (see Chapter 8). Autonomous trucks or autonomous robots have been tested for last-mile delivery of goods in urban areas. These innovative technologies may improve efficiency in terms of cost savings as well as reduce negative environmental impacts. Delivery robots are also able to provide a healthier distribution system with contactless service to customers in COVID-19 pandemic times.

Beirigo et al. (2018) studied modelling shared autonomous transportation of passengers and parcels for share-a-ride parcel locker problems in urban areas. They concluded that mixed purpose fleets for passengers and parcels performed on average 11% better than single-purpose fleets.

Boysen et al. (2018) modelled the truck-based robot delivery scheduling problem for last-mile delivery. This problem aims to find an optimal truck route and a schedule for launching autonomous robots at a stop for delivering goods to customers by minimising the weighted number of late deliveries. Simoni et al. (2020) analysed robot-assisted last-mile delivery systems and pointed out that truck-robot systems are quite efficient if robots are employed in heavily congested areas. Last-mile delivery robots allow contactless delivery to customers which can minimise the risk of infection during the COVID-19 pandemic (Chen et al., 2021). Figliozzi and Jennings (2020) showed that autonomous delivery robots have the potential for significantly reducing energy consumption and CO_2 emissions.

Scherr et al. (2019) modelled service network design for autonomous vehicles in platoons as a mixed integer liner programme. They introduced autonomous platooning in the first tier of two-tier city logistics systems. Their simulation showed that mixed autonomous fleets provide potential cost savings when deployed in city logistics.

Haasa and Friedrich (2017) studied the application of platooning in urban delivery and discussed waiting times at a transfer point and the number of platoons using microsimulation. Csiszar et al. (2018) discussed a system model for autonomous road freight transportation highlighting a structural model and an operational model focusing on system dynamics.

1.3 COLLABORATION AND COORDINATION AMONG STAKEHOLDERS

Collaboration between stakeholders is critical for successful implementation of city logistics initiatives (Taniguchi and van der Heijden, 2000). Stakeholders in city logistics include shippers, freight carriers, administrators, and residents. They have different objectives and perspectives for urban freight transport. For example, shippers and freight carriers are mainly interested in maximising their profits, while administrators try to reduce traffic congestion and global/local emissions of hazardous gases and residents are keen to ensure safety and security of communities. Therefore, they need to collaborate for planning, implementing, and evaluating city logistics policy measures.

As public entities and private companies are involved in city logistics, public-private partnerships (PPP) (Browne et al., 2014) are essential for achieving the common goals of city logistics. The first stage of PPP is identifying the problems related to urban freight transport and planning, implementing, and evaluating stages of city logistics schemes follow. In all these stages of PPP, sharing data is critical for understanding the current status, issues, and finding innovative solutions. Recently ITS, ICT, and IoT allow private logistics companies to obtain precise data on movements of vehicles and goods, location of depots, and demands for delivering goods. The public sector traditionally owns data on traffic congestion, crashes, emissions of CO_2, NOx and PM (Particulate Matter), and infrastructure provision. Data sharing within stakeholders is beneficial for recognising the issues and planning sustainable urban freight transport in the framework of public-private partnerships (Thompson et al., 2020). However, there are issues of confidentiality for logistics companies, costs of collecting data, regulations by law, and lack of tools for analysing data.

Illemann et al. (2020) discussed data sharing between private logistics companies and public policy makers and proposed a framework for identifying how shared freight data can be used to satisfy the planning needs of different freight stakeholders. Moschovou et al. (2019) pointed out that the main barriers to sharing freight data are legal barriers, resource barriers, competition barriers, and institutional barriers. Legal barriers mean the existence of laws that restrict data sharing. Resource barriers encompass the restrictions in funding and the lack of resources. Competition barriers are derived from the confidentiality of business in private companies.

Institutional barriers indicate the limitation of sharing data between various organisations.

Melo et al. (2019) performed microscopic traffic simulation modelling for discussing capacity sharing of logistics parking infrastructure owned by public authorities between private logistics companies and parents who pick up their children from school. The results showed the significant reduction in delay times, travel times, queue length, stopped times, and increase in average speeds of both private transport and freight transport. Also, considerable reductions in fuel consumption and emissions were estimated.

Discussing the effects of policy measures before implementing them as well as after their implementation is important among stakeholders in PPP. Modelling city logistics helps discussion and decision-making by providing the estimation of responses of shippers and freight carriers to policy measures taken by a municipality and their impacts on urban traffic and the environment. Katsela and Pålsson (2019) presented a multi-criteria decision model using analytic hierarchy process to support stakeholder management in city logistics.

1.4 INTEGRATED PLATFORM FOR CITY LOGISTICS

A framework for an integrated platform of city logistics is presented in Figure 1.2. Within the integrated platform, information systems in cyber space and real vehicle operations in the transport network can be integrated and multimodal freight transport is used. The integrated platform can be achieved by the support of emerging technologies of ICT, ITS, IoT, big data, AI, autonomous vehicles, and robots as well as the public-private collaboration including the use of urban consolidation centres, multimodal freight transport and data sharing (Crainic and Montreuil, 2016).

Emerging technologies allow us to take a data-driven approach. Figure 1.3 illustrates the data-driven approach for city logistics using multi-agent simulation. We can collect data on the demands of customers, vehicle types, and traffic conditions using IoT sensors and process them. Then we can analyse the behaviour of stakeholders, including shippers, freight carriers, administrators, and residents using behaviour models and AI or machine learning techniques. This analysis involves the response of logistics companies to policy measures implemented by administrators and the effects of applying autonomous vehicles for last-mile delivery. Finally, evaluation can be done on the output of vehicle movements, emissions of GHG and NOx, energy consumption, and costs.

In the case of disruption of transport networks in disasters, the integrated platform can play important roles in recognising the current status and

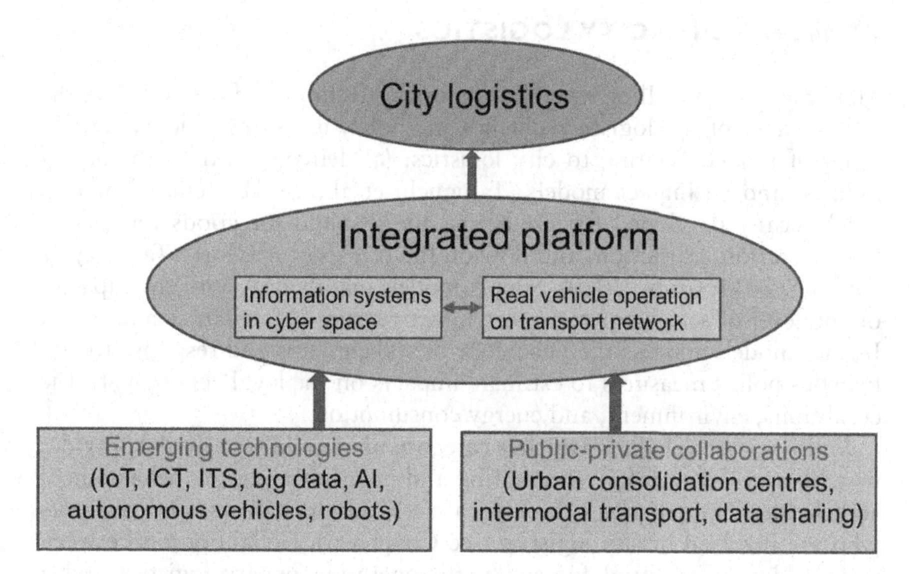

Figure 1.2 Framework of integrated platform for city logistics.

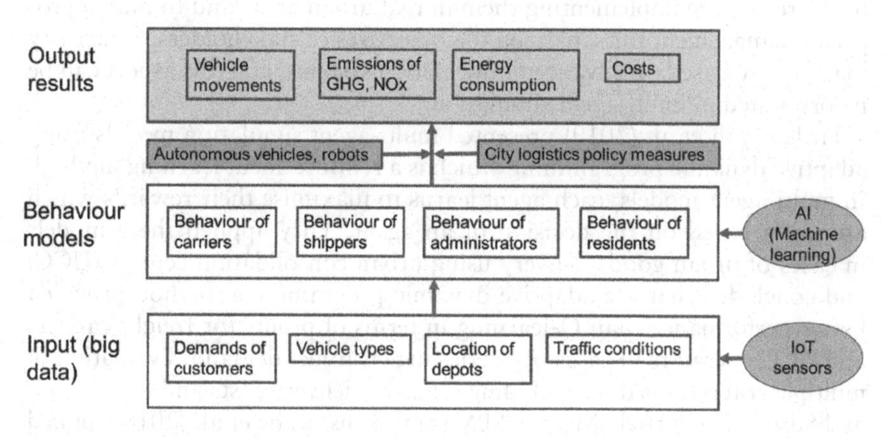

Figure 1.3 Data-driven approach for city logistics using multi-agent simulation.

finding the response solutions using IoT, ICT, ITS, big data, and AI (see Chapter 11). Quick recovery of logistics systems is critical for maintaining the services to customers. To do this, information on the road network which trucks can pass is required to find delivery routes in the damaged road network. The information can be shared with public and private players on the integrated platform. The successful operation of the integrated platform in emergency cases highly depends on the operation and training in normal cases.

1.5 MODELLING CITY LOGISTICS

Mathematical modelling provides important techniques for evaluating the performance of city logistics schemes and policy measures. There are three types of models relating to city logistics: (a) demand models, (b) supply models, and (c) impact models (Taniguchi et al., 2001). Demand models are basically developed for predicting the demand for goods relating to the generation, attraction, and flow on the transport network (Tavasszy & De Jong, 2014; Comi, 2020). Supply models mainly deal with the capacity or the level of service of transport infrastructure (Merchan et al., 2020). Impact models address the behaviour of stakeholders and response to city logistics policy measures to estimate impacts on the local economy, traffic conditions, environment, and energy consumption.

Multi-agent modelling, which is categorised as an impact model provides us with a useful tool for representing and understanding the behaviour of stakeholders, *i.e.* shippers, freight carriers, administrators, and residents who are involved in city logistics (see Chapter 5). Collaboration between stakeholders is essential for successful operation of city logistics and a multi-agent simulation allows us to predict the effects of city logistics policy measures before implementing them in real urban areas and to find appropriate management for satisfying the objectives of stakeholders. Smart city logistics are based on advanced information systems, and this aspect can be incorporated in multi-agent simulation.

Firdausiyah et al. (2019) presented multi-agent simulation models using adaptive dynamic programming which is a reinforcement learning method. In multi-agent models, each agent learns to maximise their rewards which are given based on the decision of an agent. They applied these models in cases of urban goods delivery using urban consolidation centres (UCC) and concluded that the adaptive dynamic programming method provided better performance than Q-learning in terms of profits for freight carriers and UCC operators in cases of fluctuating customer demand. Evaluation of multiple criteria was done including reducing delivery costs and CO_2, NOx, and Suspended Particle Matter (SPM) emissions. Rabe et al. (2018) applied a discrete event simulation to evaluate a UCC for distribution in Athens. They concluded that the simulation showed that utilising UCCs lowered the number of tours needed to meet the customer's demands and decreased the required distance travelled by freight vehicles.

Joubert (2018) applied the multi-agent transport simulation model MATSim for evaluating the relocation of an urban container terminal in large-scale scenarios in Cape Town, South Africa. Results showed that the simulation provided insights about how specific operators and the public at large will be influenced, both from an overall vehicle-kilometres travelled point of view as well as a change in travel times.

Jlassi et al. (2017) evaluated regulatory policies including vehicle size and time windows in city logistics using a multi-agent simulation in Paris.

They tested four scenarios and discussed the simulation results from various viewpoints including the number of vehicles, emissions of CO_2, NOx, and SO_2.

Taniguchi et al. (2018b) demonstrated how multi-agent simulation can be used for evaluating a combination of city logistics policy measures focusing on consolidation centres, green management, and parking management. They pointed out that in public-private partnerships (PPP), multi-agent models will play an important role in evaluating city logistics policy measures based on data. Sharing data from both the public and private sectors is essential, and interaction between both sectors is necessary to find better solutions using the results provided by multi-agent simulation.

Multi-agent models can be used for multi-stakeholder decision-making based on this modelling approach. Le Pira et al. (2017) applied integrated models of discrete choice models (DCM) with agent-based models (ABM). These models consider stakeholders' heterogeneous preferences and simulate their interactive behaviour in a consensus-building process. They pointed out that integrating DCM and ABM can overcome their weaknesses since it is grounded on sound microeconomic theory providing a detailed (static) stakeholders' behavioural knowledge, but it is also capable of reproducing agents' (dynamic) interaction during the decision-making process. Lebeau et al. (2018) presented the multi-actor multi-criteria analysis (MAMCA) as a method which structured the consultation process of stakeholders in city logistics (see Chapter 7). They tested this approach in a workshop in Brussels and concluded that the methodology allowed authorities to identify the priorities of stakeholders in city logistics, to guide the discussion towards a consensus and to provide inputs for an eventual adoption of strategies. Gatta et al. (2019) applied an interactive MAMCA for analysing off-hour delivery (OHD) solutions with stakeholders in Rome. They pointed out that stakeholders prefer a solution where OHD is jointly implemented together with one or more urban consolidation centres.

The Vehicle Routing and Scheduling Problem with Time Windows (VRPTW) models (Solomon, 1987; Taniguchi & Shimamoto, 2004; Braysy & Gendreau, 2005; Qureshi et al., 2012; Schneider, 2016; Bettinelli et al., 2019) have been used to represent the behaviour of freight carriers in urban distribution (see Chapter 4). In these models, pickup-delivery trucks visit multiple customers using a single route, and time windows are commonly set by customers in Just-In-Time transport situation. VRPTW models are suitable for mathematically representing the behaviour of freight carriers. Stochastic and dynamic VRPTW models are usually used in real-time operation of vehicles to incorporate varying travel times and demands of customers. Stochastic and dynamic VRPTW models (Laporte et al., 1992; Kenyon et al., 2003; Zhang et al., 2013; Ehmke et al., 2015) can be applied with historical and real-time travel times data on road networks using ITS. Multi-agent simulation models can incorporate VRPTW models as sub-models to represent the route choice behaviour of freight carriers.

1.6 CONCLUSIONS

City Logistics provides innovative solutions for urban freight transport issues using emerging technologies such as Information and Communication Technologies (ICT), Intelligent Transport Systems (ITS), Internet of Things (IoT), big data, Artificial Intelligence (AI), autonomous vehicles, and robots. Collaboration between stakeholders, such as shippers, freight carriers, administrators, and residents, is critical for successful implementation of city logistics initiatives. The integrated platform can promote city logistics. Within the integrated platform, information systems in cyberspace and real vehicle operations in the transport network can be integrated and multi-modal freight transport is used. Mathematical modelling provides important techniques for evaluating the performance of city logistics schemes and policy measures.

REFERENCES

Beirigo, B. A., Frederik Schulte, F. and Negenborn, R. R. (2018). Integrating people and freight transportation using shared autonomous vehicles with compartments, *IFAC Papers OnLine*, **51–59**, 392–397.

Bettinelli, A., Cacchiani, V., Crainic, T.G., Vigo, D. (2019). A branch-and-cut-and-price algorithm for the multi-trip separate pickup and delivery problem with time windows at customers and facilities, *European Journal of Operational Research*, **279**, 824–839.

Bibri, S.E. (2018). The IoT for smart sustainable cities of the future: An analytical framework, *Sustainable Cities and Society*, **38**, 230–253.

Boysen, N., Stefan Schwerdfeger, S. and Weidinger, F. (2018). Scheduling last-mile deliveries with truck-based autonomous robots, *European Journal of Operational Research*, **271**, 1085–1099.

Braysy, O. and Gendreau, M. (2005). Vehicle Routing Problem with Time Windows, Part II: Metaheuristics, *Transportation Science*, **39(1)**, 119–139.

Browne, M., Lindholm, M. and Allen, J. (2014). Partnerships among stakeholders, In: Taniguchi, E. and Thompson, R.G. (eds.), *City Logistics: Mapping the Future*, CRC Press, New York, 13–24.

Chen, C., Demir, E., Huang, Y. and Qiu, R. (2021). The adoption of self-driving delivery robots in last mile logistics, *Transportation Research Part E*, **146**, 102214.

Cleophas, C., Cottrill, C., Jan Fabian Ehmke, J. F. and Tierney, K. (2019). Collaborative urban transportation: Recent advances in theory and practice, *European Journal of Operational Research*, **273**, 801–816.

Comi, A. (2020). A modelling framework to forecast urban goods flows, *Research in Transportation Economics*, **80**, 100827.

Crainic, T.G. and Montreuil, B. (2016). Physical internet enabled hyperconnected city logistics, *Transportation Research Procedia*, **12**, 383–398.

Csiszar, B. and Foldes, D. (2018). System model for autonomous road freight transportation, *Promet – Traffic & Transportation*, **30**, 93–103.

Dupas, R., Taniguchi, E., Deschamps, J-C., Qureshi, A. (2020). A multi-commodity network flow model for sustainable performance evaluation in city logistics: Application to the distribution of multi-tenant buildings in Tokyo, *Sustainability*, **12**, 2180.

Ehmke, J.F., Campbell, A.M., Urban, T.L. (2015). Ensuring service levels in routing problems with time windows and stochastic travel times, *European Journal of Operational Research*, **240**, 539–550.

Figliozzi, M. and Jennings, D. (2020). Autonomous delivery robots and their potential impacts on urban freight energy consumption and emissions, *Transportation Research Procedia*, **46**, 21–28.

Firdausiyah, N., Taniguchi, E. and Qureshi, A.G. (2019). Modeling city logistics using adaptive dynamic programming based multi-agent simulation, *Transportation Research Part E*, **125**, 74–96.

Gatta, V., Marcucci, E., Site, P. D., Le Pira, M. and Carrocci, C. S. (2019). Planning with stakeholders: Analysing alternative off-hour delivery solutions via an interactive multi-criteria approach, *Research in Transportation Economics*, **73**, 53–62.

Haasa, I. and Friedrich, B. (2017). Developing a micro-simulation tool for autonomous connected vehicle platoons used in City Logistics. *Transportation Research Procedia*, **27**, 1203–1210.

Illemann, T.M., Karam, A. and Reinau, K.H. (2020). Towards sharing data of private freight companies with public policy makers: A proposed framework for identifying uses of the shared data, *Proceedings of the 2020th IEEE 7th International Conference on Industrial Engineering and Applications (ICIEA)*, Bangkok, Thailand, 16–21 April 2020.

Jlassi, S., Tamayo, S., Gaudron, A. and de La Fortelle, A. (2017). Simulating impacts of regulatory policies on urban freight: application to the catering setting, *6th IEEE International Conference on Advanced Logistics and Transport (ICALT)*, 106–112.

Joubert, J. (2018). Evaluating the relocation of an urban container terminal, In: E. Taniguchi and R. G. Thompson (Eds.), *City Logistics* 2, ISTE-Wiley, London, 197–210.

Katsela, K. and Pålsson, H. (2019). A multi-criteria decision model for stakeholder management in city logistics, *Research in Transportation Business & Management*, **33**, 100439.

Kenyon, A.S., Morton, D.P. (2003). Stochastic vehicle routing with random travel times. *Transportation Science*, **37**, 69–82.

Laporte, G., Louveaux, F., Mercure, H. (1992). The vehicle routing problem with stochastic travel times, *Transportation Science*, **26**, 161–170.

Le Pira, M., Marcucci, E., Gatta, V., Ignaccolo1, M., Inturri, G. and Pluchino, A. (2017). Towards a decision-support procedure to foster stakeholder involvement and acceptability of urban freight transport policies, *European Transportation Research Review*, **9**, 54.

Lebeau, P., Cathy Macharis, C., Van Mierlo, J. and Janjevic, M. (2018). Improving policy support in city logistics: The contributions of a multi-actor multi-criteria analysis, *Case Studies on Transport Policy*, **6**, 554–563.

Mehmood, R., Meriton, R., Graham, G., Hennelly, P. and Kumar, M. (2017). Exploring the influence of big data on city transport operations: A Markovian approach, *International Journal of Operations & Production Management*, **37**, 75–104.

Melo, S., Macedo, J., Patrícia Baptista, P. (2019). Capacity-sharing in logistics solutions: a new pathway towards sustainability, *Transport Policy*, 73, 143–151.

Merchan, D., Winkenbach, M., Snoeck, A. (2020). Quantifying the impact of urban road networks on the efficiency of local trips, *Transportation Research Part A*, **135**, 38–62.

Moschovou, T., Vlahogianni, E. I. and Rentsiou, A. (2019). Challenges for data sharing in freight transport, *Advances in Transportation Studies: An International Journal Section*, **B48**, 141–152.

Qureshi, A.G., Taniguchi, E. and Yamda, T. (2012). An analysis of exact VRPTW solutions on ITS data-based logistics instances, *International Journal of ITS Research*, **10**, 34–46.

Rabe, M., Klueter, A. and Wuttke, A. (2018). Evaluating the consolidation of distribution flows using a discrete event supply chain simulation tool: Application to a case study in Greece, *Proceedings of the IEEE 2018 Winter Simulation Conference*, Gothenburg, Sweden.

Scherr, Y. O., Saavedra, B. A. N., Hewitt, M. and Mattfeld, D. C. (2019). Service network design with mixed autonomous fleets, *Transportation Research Part E*, **124**, 40–55.

Schneider, M. (2016). The vehicle -routing problem with time windows and driver-specific times, *European Journal of Operational Research*, **250**, 101–119.

Simoni, M.D., Erhan Kutanoglu, E. and Claudel, C.G. (2020). Optimization and analysis of a robot-assisted last mile delivery system, *Transportation Research Part E*, **142**, 102049.

Solomon, M.M. (1987). Algorithms for the vehicle routing and scheduling problems with time window constraints. *Operations Research*, 35, 254–265.

Taniguchi, E. and Shimamoto, H. (2004). Intelligent transportation system based dynamic vehicle routing and scheduling with variable travel times, *Transportation Research Part C*, **12C**, 235–250.

Taniguchi, E. and Thompson, R. G. (2014). *City Logistics: Mapping the Future*, CRC Press, New York.

Taniguchi, E. and van der Heijden, R.E.C.M. (2000). An evaluation methodology for city logistics. *Transport Reviews*, **20**, 65–90.

Taniguchi, E., Dupas, R, Deschanmps, J-C. and Qureshi, A. G. (2018a). Concepts of an integrated platform for Innovative city logistics with urban consolidation centers and transshipment points, In: Taniguchi, E. and Thompson, R.G. (eds), *City Logistics 3: Towards Sustainable and Liveable Cities*, ISTE, London, 129–146.

Taniguchi, E., Qureshi, A. G. and Konda, K. (2018b). Multi-agent simulation with reinforcement learning for evaluating a combination of city logistics policy measures, In: Taniguchi, E. and Thompson, R.G. (eds), *City Logistics 2: Modelling and Planning Initiatives*, ISTE, London, 165–178.

Taniguchi, E., Thompson, R. G., Yamada, T. and van Duin, R. (2001). *City Logistics – Network Modelling and Intelligent Transport Systems*. Elsevier, Pergamon, Oxford.

Taniguchi, E., Thompson, R.G. and Qureshi, A.G. (2020). Modelling city logistics using recent innovative technologies, *Transportation Research Procedia*, **46**, 3–12.

Tavasszy, L. and De Jong, G. (2014). *Modelling Freight Transport*, Elsevier, Oxford.
Thompson, R.G., Nassir, N. and Frauenfelder, P. (2020). Shared freight networks in metropolitan areas, *Transportation Research Procedia*, **46**, 204–211.
Zhang, J., William H. K. Lam, W.H.K., Bi Yu Chen, B.Y. (2013). A stochastic vehicle routing problem with travel time uncertainty: Trade-off between cost and customer service, *Networks and Spatial Economics*, **13**, 471–496.

Chapter 2

Data collection and analyses

2.1 DATA COLLECTION USING IoT, ICT, AND ITS

Analysis in city logistics often involves working with vehicle routing problems, which are defined on a complete graph (*i.e.*, a fully connected network) where customers are represented as vertices and are interconnected with direct arcs. Most of the related research also bases their analysis on such complete graphs, for example, benchmark instances developed by Solomon (1987). In real-life logistics instances, each arc of the complete graph represents a combination of actual urban road network links, which formulate the shortest path between the origin and destination nodes (*i.e.*, two vertices of the complete graphs). Advancements in Intelligent Transportation Systems (ITS) enable the collection of a large amount of useful road network-related data. It is advantageous to base city logistics analysis on realistic datasets using actual traffic data as it enables the option of comparing other solution characteristics (except the solution cost and the number of vehicles) typically based on actual traffic conditions. For example, generation of various emissions can be considered, which depends on the travel speed. Adopting real road networks for logistics instances is a challenge as it may involve a large number of links. Collecting the actual traffic-related data such as travel time and speed, could be a time consuming and cumbersome process without the use of ITS. The accuracy and tractability of ITS data also provides additional advantages. Another benefit of using actual road networks and associated traffic conditions is that truck routes can be mapped to road links, which can help monitor trucks' movement as well as identify links with severe problems commonly attributed to freight truck movement. For example, truck emissions can be calculated and mapped on each link of the road network. Such emission maps can be used to identify links, which require some intervention from public or private side to abate severe environmental issues.

In following sections, some examples of data collection based on ITS and ICT applications in the field of city logistics are presented.

DOI: 10.1201/9781003261513-3

2.1.1 VICS data

The Vehicle Identification and Communication System (VICS) is one of the important ITS applications, which provides useful traffic information to users in order to relax the congestion on the roads. The first VICS in the world commenced its service in Japan on 23 April 1996 (Yamada, 1996). It was designed to distribute traffic data collected by prefectural police headquarters and road administrators, using the roadside infrastructure consisting of radio wave beacons, infrared beacons, and frequency-modulated (FM) multiplex broadcasting. The information provision was achieved using maps, simple graphics and text messages. VICS proved to be very successful in Japan, and by the end of the year 2009 about 25.9 million VICS units have been sold in Japan, which now come as a standard in car navigation systems. At present, VICS data is augmented with a probe car system, another important ITS application in Japan.

Figure 2.1 shows a sample of VICS data for links to road network. It shows the travel time on a particular link every 5 minutes. Figure 2.2 shows the variations of travel time on a road network link recorded on a particular day from VICS data. A large number of road network links are involved in city logistics analysis. Considering detailed travel time variations (shown in Figure 2.2(a)) or dynamic processing of VICS data as it arrives can be very challenging. Instead, researchers have used distributions obtained from VICS data (such as shown in Figure 2.2(b)) to cover travel time variations. Static versions of vehicle routing can be solved using only average data (over the complete or partial time period) presented in Figure 2.2(a) (as in Qureshi et al., 2012). A time step function can also be developed by considering averages over multiple time periods as shown in Figure 2.3. Fleischmann et al. (2004) worked with test instances based on such time stepwise travel time data

1		Link											
2	Time	1	2	3	4	5	6	7	8	9	10	11	12
3	0	133	159	101	84	130	86	97	69	60	54	209	187
4	5	108	172	112	125	118	84	88	91	67	57	209	223
5	10	113	182	123	149	120	84	88	105	74	60	240	220
6	15	133	181	114	159	129	87	94	111	65	65	251	203
7	20	142	183	107	152	129	88	94	106	57	64	244	202
8	25	143	179	111	141	118	90	88	100	60	60	237	201
9	30	145	170	112	139	118	94	88	101	62	60	222	187
10	35	147	170	112	141	127	91	94	102	60	63	218	186
11	40	146	166	115	142	130	88	94	102	62	60	230	185
12	45	130	166	115	147	122	91	89	102	63	59	238	187
13	50	122	179	116	152	120	93	89	105	65	59	241	193
14	55	137	184	115	145	128	92	96	103	68	54	251	181
15	60	144	171	105	139	125	92	95	101	70	58	253	195
16	65	136	163	100	145	116	95	88	102	65	69	240	233
17	70	137	168	104	153	115	103	87	107	60	70	238	229
18	75	141	173	106	152	125	102	93	107	62	67	246	212
19	80	148	179	111	148	129	94	95	106	63	72	246	217
20	85	148	178	112	149	123	93	91	107	62	70	243	206
21	90	150	174	112	150	122	96	92	107	63	64	245	199

Figure 2.1 Sample of travel time data provided by VICS.

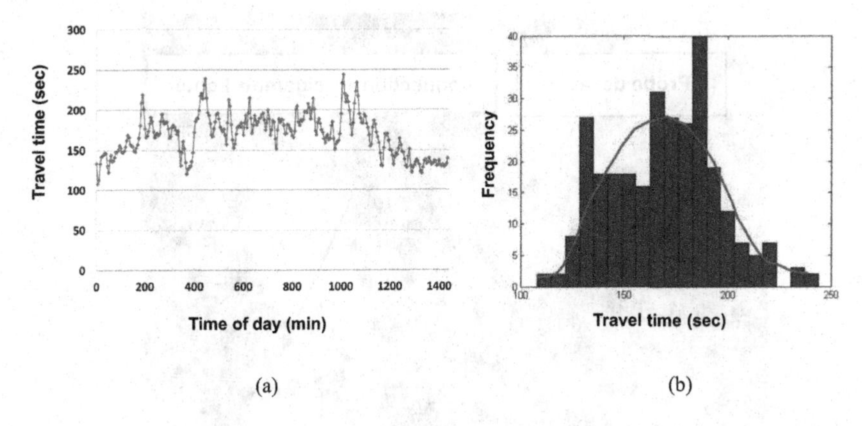

Figure 2.2 Travel time variation on a link in VICS data. (a) Raw data. (b) Histogram and distribution.

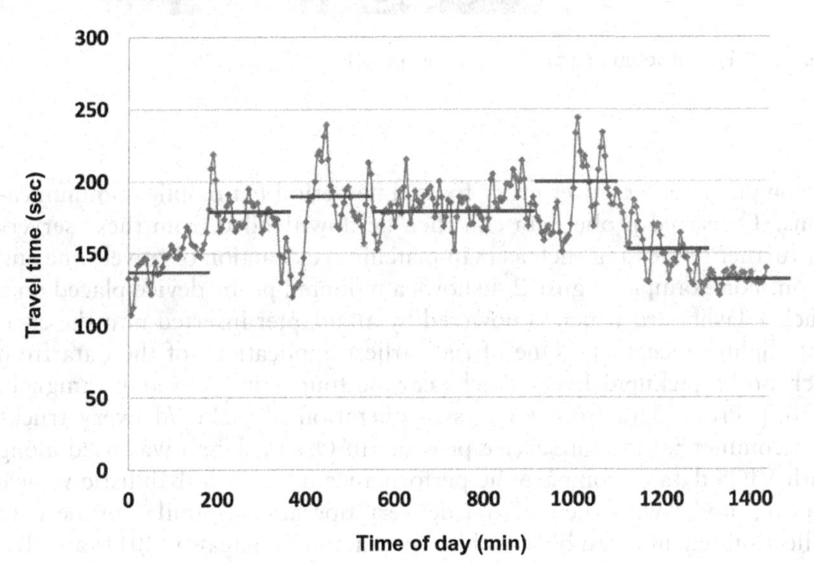

Figure 2.3 Stepwise travel time on a link based on VICS data.

based on an ITS system implemented in Berlin, Germany that also provides travel time data on links in 5-minute slots. They also provided a smoothening technique to avoid problems while working with stepwise travel time data.

2.1.2 Probe data

Travel time data can be obtained by using communication with a global positioning system (GPS) device mounted in the vehicles called probe vehicles. Travel location (latitude and longitude) of these probe vehicles is sent

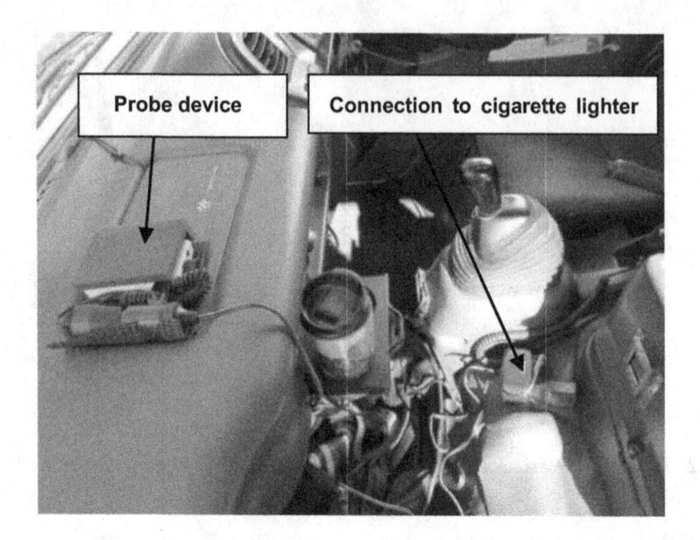

Figure 2.4 Typical setup of a probe device in a truck.

to computer servers after every fixed time period *via* mobile communications. The saved probe data can then be downloaded from these servers for further processing such as map matching, extraction of travel time and so on. For example, Figure 2.4 shows a mounted probe device placed on a truck's dashboard which is powered by an adapter inserted into the cigarette lighter receptacle. One of the earliest applications of the data from such probe pickup/delivery trucks can be found in Ando and Taniguchi (2006). Probe data from 66 days of operation of pickup/delivery trucks of a commercial freight service provider in Osaka, Japan was used along with VICS data to compare the performance of their probabilistic vehicle routing model with their actual delivery operations. Similar probe data collection schemes can be found in Yokota and Tamagawa (2011) and Teo et al. (2015).

Although probe data provides a rich resource of location and travel time data of vehicles, a lot of other data is required to analyse the urban freight, such as customers' location and the delivery quantity (*i.e.*, customers' demand) as well as time windows. Freight carriers are usually very reluctant to share this type of data due to privacy and other business-related issues. Some researchers (such as Teo et al., 2015) have tried to extract this data from probe data but with limited success. In order to fill in some of the above-mentioned data gaps, Teo et al. (2015) augmented probe data collection with simultaneous manual data collection, in which surveyors followed the probe trucks and recorded additional data.

2.2 BIG DATA ANALYSES IN URBAN FREIGHT

Systems such as VICS and probe data provide big data for city logistics. Their analysis can extract a lot of valuable information for improvements in city logistics systems. For example, using a logistics instance based on VICS data, Qureshi et al. (2012) mapped the CO_2 emissions (which depend on travel speed) on the links used by delivery vehicles as well as for their idling stops. Their analysis shows that considering hard time windows in vehicle routing produces more severe environmental problems than the case when time windows are considered flexible (*i.e.*, soft time windows).

Based on analysis of probe data of 300 delivery trucks, Yokota and Tamagawa (2011) found that the road network in the Keihanshin area (Osaka and Kobe, Japan) can be divided into two types of networks: the high-frequency network (HFN) and the low-frequency network (LFN). The HFN is only a quarter of the size of the LFN, but delivery trucks use it about six times more than the LFN. The average speed on the HFN was found to be higher as well, *i.e.*, about 25 km/h as compared to only 17 km/h for the LFN. Yokota and Tamagawa (2011) suggested using mainly the HFN for planning purposes for longer delivery routes whereas the LFN should be used for access and egress (first and the last mile) only. It was further reported that delivery trucks seldom use urban motorways or toll highways for routes smaller than 20 km; therefore, these can be ignored in the analysis of intra-city movement.

Using a probe dataset based on the delivery trucks of a commercial freight carrier in Osaka, Japan and the land-use maps, Qureshi et al. (2018) found that a considerable movement of the delivery trucks occurs in the residential areas, which can be avoided if such movements are penalised in the objective function.

2.3 DATA SHARING AMONG STAKEHOLDERS

2.3.1 Overview

Data sharing between the public sector and private companies is important for modelling, planning and evaluating policy measures related to city logistics. Recently ITS (Intelligent Transport Systems), ICT (Information and Communication Technologies), and IoT (Internet of Things) allows private logistics companies to obtain precise data on movements of vehicles and goods, location of depots and demands for delivering goods. Public sector traditionally owns data on traffic congestion, crashes, emissions of CO_2, NOx, and PM, and infrastructure provision. Data sharing among stakeholders is beneficial for understanding the current status of freight networks and planning sustainable urban freight transport in the framework of public-private partnerships. However, there are issues for logistics companies, such as confidentiality, costs of collecting data, regulations by law and lack of tools for analysing data.

Data sharing between the public sector and private companies is essential for understanding the current status of urban freight transport. National and local governments try to encourage data sharing in urban freight transport. However, there are issues and barriers in for sharing in practice. Moschovou et al. (2019) pointed out that the main barriers met in sharing freight data are legal barriers, resource barriers, competition barriers, and institutional barriers. Legal barriers involve laws that restrict data sharing. Resource barriers encompass restrictions in funding and the lack of resources. Competition barriers are derived from the confidentiality of business in private companies. Institutional barriers indicate the limitation of sharing data between various organisations. In spite of these issues and barriers, some good examples of freight data sharing are observed in several countries in Europe as well as Japan and Australia, which are highlighted below.

2.3.2 Examples of data sharing

In Rotterdam, the Netherlands, data sharing for improving urban freight transport is performed among the City of Rotterdam, 56 private companies, and two universities (van Duin, 2020). A covenant for sharing data between the City of Rotterdam and 56 companies that included receivers, transporters, shippers, retailers, financial service providers, vehicle manufacturers and a think tank was signed in 2020. Universities take part in carrying out urban freight flow simulations using the data provided by private companies.

In Japan the Ministry of Land, Infrastructure, Transport and Tourism (MLIT, 2010) collects probe data of vehicle movements using ETC2.0 (Electric Toll Collection 2.0) on-board devices and ITS Spot roadside units (Taniguchi, 2020). These data include the traffic of passenger cars and freight vehicles. The data collected from freight vehicles can be used by logistics companies and applied to vehicle management and improving driving safety. This initiative started in 2018. Benefits for logistics companies using ETC2.0 data include (a) less expensive compared to the case of setting up on-board devices and communication systems by themselves and (b) standardised data format is already given by the data provider and ETC2.0 data is easy to use for vehicle management. Truck routing data by ETC2.0 can be used for optimising truck routing and scheduling as well as reducing emissions of CO_2, NOx, and SPM (suspended particulate matter).

In Japan the framework for data sharing between the public sector and private companies is given by Basic Act on the advancement of public and private sector data utilisation (2016). The Act determines the responsibilities of the state, local public entities and private companies by providing basic principles with respect to the advancement of the appropriate and effective use of public and private sector data.

In France, there are some examples of public-private exchange of information for public interest in the City of Paris, Lyon and Saint-Etienne. In Lyon-Saint-Etienne area, two public bodies and two private companies and four research units participated in the ELUD (Efficacité de la logistique urbaine alimentaire durable) project (research based) (Gonzalez-Feliu, 2020). They carried out route data sharing, spatial data sharing, use of standard research-based demand and supply estimation tools. Some results supported public and private decision making (Palacios-Arguello et al., 2018; Gonzalez-Feliu, 2019). Open data are also available and spatial and functional data are used to support public-private collaboration in decision making *via* a web-based tool (Gonzalez-Feliu, 2017).

In Australia, Bureau of Infrastructure, Transport & Regional Economics (BITRE) made Non-Disclosure Agreement with a number of transport companies and obtained truck GPS (Global Positioning System) data (Thompson, 2020) (See Chapter 3). The government uses these data for understanding traffic congestion and for improving traffic management and infrastructure provision.

In the UK data sharing and information sharing is carried out within Freight Quality Partnerships for urban freight transport (Wainwright, 2020). For example, the Central London Freight Quality Partnership (CLFQP) with Transport for London (TfL) and private sector organisations started in 2006 for central London. TfL executed street space initiatives which aim at creating more space for people to safely walk, cycle, scoot, or ride in response to the COVID-19 pandemic in 2020. Information on the change in road space usage, including street closures, footways made wider, restricted through traffic is shared with private companies, since this type of information is useful for the smooth delivery of goods.

In Sweden sharing data, knowledge and visions is performed between public and private sectors. For example, in Gothenburg, some industry and public data are combined for a better understanding and planning of transport infrastructure (Flodén et al., 2020).

2.3.3 Discussion

There is a large variety of data sharing between the public and private sectors for urban freight transport in different countries due to laws, cultures, and businesses environment.

Collaboration in data sharing is beneficial for both public and private sectors in the context of public-private partnerships in urban freight transport. For the public sector, they can utilise real data of logistics activities of private companies for developing policies on urban freight transport. For private companies they can enjoy lower costs for data collection and usage. If data is provided in a standardised format, they can easily use the data for collaborative delivery with other companies.

Regarding the confidentiality of private companies' data, a covenant between public sectors and private companies can be effective to allow the public sector to use a company's data but not disclose it to other competitive companies. This type of covenant is needed to ensure the confidentiality of private companies. If some companies agree to operate joint delivery systems, for example, it is welcome to use data of multiple companies who take part in joint delivery systems. Therefore, the clear agreement and covenant can promote reliable and effective initiatives.

In Sweden, when private data are given to the public sector, the data may be legally recognised as "official records" which means it is accessible to the public. This may be a reason for private companies not being willing to be involved in data sharing with the public sector. Shared data should be under a trusted party and the public authority may not be a trusted party, if the public authority needs to disclose the shared data coming from private companies to the public.

Data sharing between the public and private sectors is usually undertaken using volunteer-based schemes. In certain cases in the Netherlands, Japan and Australia volunteer-based data sharing has been successfully carried out. Incentives for private companies to be involved in data sharing with the public sector and other private companies may be lower costs for collecting data, access to better software for analysing data, and use of standardised data formats.

In city logistics data sharing is a part of public-private partnerships (PPP) between stakeholders. Therefore, based on data sharing, the stakeholders are able to understand the current conditions of urban freight transport networks, analyse policy measures, implement them and evaluate their effects. The total activities in PPP should be well coordinated. In this sense, the limitation of data affects the analyses and results and then sharing data can overcome the limitation of data of a single entity.

Regarding the quality of data, some institutions should guarantee the quality of data. The standardisation of the method for collecting and processing data is critical for analyses. Usually, we face missing spatial and temporal data in urban freight transport. How extrapolation can be performed with missing data is an important problem for obtaining good results with higher accuracy.

In the time of the COVID-19 pandemic, data sharing is important between stakeholders, since the worldwide pandemic is the first time we have faced the difficulty of estimating the sudden change of demand for customers. Also, as there is uncertainty in estimating the behaviour of customers in the pandemic, we need to share data and experience associated with urban deliveries.

2.3.4 Blockchain

The concept of blockchain was first introduced by Nakamoto (2008) for the cryptocurrency Bitcoin. Blockchain is distributed, consensus-based and (mostly) immutable ledger of transaction records (Schmidt & Wagner,

2019). Blockchain technology can be used to realise data sharing and build trust among stakeholders of city logistics in the cooperative delivery of goods in urban areas (Hribernik et al., 2020; Li et al., 2022). The blockchain adopts a peer-to-peer mechanism to ensure that the data in the distribution process is immutable, which does not require a third party to secure the information. However, the research and practices of the application of blockchain in city logistics is still limited.

2.4 CONCLUSIONS

Advancements in Intelligent Transportation Systems (ITS), Information and Communication Technology (ICT), and Internet of Things (IoT) enable the collection of a large amount of useful road network-related data. These big data in real-time or historical domain can help analyse urban freight transport systems using Artificial Intelligence (AI) to find innovative solutions for city logistics. Data sharing between the public sector and private companies is important for modelling, planning and evaluating policy measures related to city logistics. Although there are some issues and barriers to achieve data sharing, some good examples of freight data sharing have been observed in several countries in Europe as well as Japan and Australia. Blockchain technology can be used to realise data sharing and build trust among stakeholders of city logistics for cooperative delivery of goods in urban areas.

REFERENCES

Ando, N., and Taniguchi, E. (2006). Travel time reliability in vehicle routing and scheduling with time windows, *Networks and Spatial Economics*, 6, 293–311.

van Duin, R. (2020). Policies in Rotterdam, In: *Workshop on Data Sharing of Public Sectors and Private Companies for Developing Policies in City Logistics*, FFJ/EHESS, Paris.

Fleischmann, B., Gietz, M. and Gnutzmann, S. (2004). Time-varying travel times in vehicle routing", *Transportation Science*, 38, 160–173.

Flodén, J., Bäckman, T., Nuldén, U., and Woxenius, J. (2020). Data sharing between public sectors and private companies, In: *Workshop on Data Dharing of Public Sectors and Private Companies for Developing Policies in City Logistics*, FFJ/EHESS, Paris.

Gonzalez-Feliu, J. (2017). Aide à la décision pour le développement de schémas logistiques urbains durables, *Presentation made at the Stand of ANNONA Project at Innovative SHS 2017, Marseille, France.*

Gonzalez-Feliu, J. (2019). Project ELUD: Efficacité de la logistique urbaine alimentaire durable. Pitch proposed for Solutrans workshop on innovative projects in transportation, CARA stand, Solutrans, Lyon, France, November.

Gonzalez-Feliu, J. (2020). Case study: France, Public open data, data sharing/collaboration practices and the role of research in promoting them, In: *Workshop on*

Data Sharing of Public Sectors and Private Companies for Developing Policies in City Logistics, FFJ/EHESS, Paris.

Hribernik, M., Zero, K., Kummer, S. and Herold, D.M. (2020). City logistics: Towards a blockchain decision framework for collaborative parcel deliveries in micro-hubs. *Transportation Research Interdisciplinary Perspectives*, 8, 100274.

Li, Y., Lin, M.K. and Wang, C. (2022). An intelligent model of green urban distribution in the blockchain environment, *Resources, Conservation & Recycling*, **176**, 105925.

Ministry of Land, Infrastructure and Transportation (MLIT) web site. Available at http://www.mlit.go.jp/road/ITS/topindex /topindex_g03_3.html (accessed April 15, 2010).

Moschovou, T., Vlahogianni, E. I. and Rentsiou, A. (2019). Challenges for data sharing in freight transport, *Advances in Transportation Studies: An International Journal Section*, **B48**, 141–152.

Nakamoto, M. (2008). *Bitcoin: A Peer-to-Peer Electronic Cash System*, Satoshi Nakamoto Institute. Available at https://satoshi.nakamotoinstitute.org

Palacios-Argüello, L., Gonzalez-Feliu, J., Gondran, N., and Badeig, F. (2018). Assessing the economic and environmental impacts of urban food systems for public school canteens: Case study of Great Lyon Region. *European Transport Research Review*, **10**, 37.

Qureshi, A.G., Taniguchi, E. and Yamada, T. (2012). An analysis of exact VRPTW solutions on ITS data-based logistics instances, *International Journal of Intelligent Transport Systems Research*, **10**, 34–46.

Qureshi, A. G., Taniguchi, E., Mai, V.P. and Teo, J.S.E. (2018). Sustainable city logistics - managing land use footprint of last-mile freight activities, *Transport Research Board 97th Annual Meeting*, Washington, DC.

Schmidt, C.G. and Wagner, S.M. (2019). Blockchain and supply chain relations: A transaction cost theory perspective, *Journal of Purchasing and Supply Management*, **25**, 100552.

Solomon, M.M. (1987). Algorithms for the vehicle routing and scheduling problems with time window constraints. *Operations Research*, **35**, 254–265.

Taniguchi, E. (2020). Data sharing between public sectors and private companies in Japan, In: *Workshop on Data Sharing of Public Sectors and Private Companies for Developing Policies in City Logistics*, FFJ/EHESS, Paris.

Teo, J.S.E., Taniguchi, E., Qureshi, A.G., Mai, V.P., Uchiyama, N. (2015). Towards a safer and healthier urbanization by improving land use footprint of last-mile freight delivery, *Transportation Research Board 94th Annual Meeting*, Washington, DC.

Thompson, R. G. (2020). Urban freight data sharing - Australian experiences, In: *Workshop on Data Sharing of Public Sectors and Private Companies for Developing Policies in City Logistics*, FFJ/EHESS, Paris.

Wainwright, I. (2020). Data sharing between public and private sectors, In: *Workshop on Data Sharing of Public Sectors and Private Companies for Developing Policies in City Logistics*, FFJ/EHESS, Paris.

Yamada, S. (1996). The strategy and deployment plan for VICS, *IEEE Communications Magazine*, **34**, 94–97.

Yokota, T. and Tamagawa, D. (2011). Constructing two-layered freight traffic network model from truck probe data, *International Journal of ITS Research*, **9**, 1–11.

Chapter 3

Geographic information systems and spatial analysis

3.1 INTRODUCTION

Recently, there has been considerable developments in spatial analysis tools that can be used to create information that can be used by decision makers to enhance the performance of urban freight systems. Global Position Systems (GPS) offer an effective means of logging the movement of freight vehicles allowing analysis of routes as well as their performance to be undertaken. Geographic Information Systems (GIS) provide an effective means of mapping a range of freight related data that can provide information that links various stakeholder interests. There are a number of tools available for understanding the spatial patterns of freight demand and impacts within urban areas. Exposure of residents to impacts from carriers can be analysed. The performance of segments of freight networks for carriers can be displayed allowing infrastructure planners and managers to identify bottlenecks.

3.2 GLOBAL POSITION SYSTEMS

3.2.1 Introduction

Global Position Systems (GPS) can provide a continuous log of data relating to the temporal location and speed of freight vehicles in urban areas providing a low cost, automated, passive and unobtrusive means of monitoring the movement of freight vehicles. This section illustrates how GPS data can be processed to create information on the performance of routes and road networks as well as freight demand patterns in cities.

3.2.2 Vehicle routes

GPS data provides a useful resource for analysing freight vehicle routes. It can potentially eliminate manual surveys of vehicle activities

DOI: 10.1201/9781003261513-4

(Toilier & Gardrat, 2019). GPS has been used to replace manual surveys allowing a wide range of analytics to be conducted (Alho, 2018a).

A number of performance metrics can be defined and calculated to investigate the efficiency of delivery routes as well as compare performance for different periods of the day. There is a need to estimate and analyse characteristics of vehicle routes including number of deliveries, stopped time, and running time (time vehicle is in motion) (Fransoo, et al., 2019; Sanchez-Diaz et al., 2019). Efficiency of driving (average speed), reliability (standard deviation of travel time) and service (service time and number of service stops per hour) are also important (Brenden et al., 2017).

Determining stops for unloading and loading goods can be challenging since there are a number of other activities that involve vehicle stops such as intersections, driver breaks and refuelling (Siripirote et al., 2020).

Tour analysis involves estimating the number of mobility-related performance measures such as travel time between stops, service time, and travel time between delivery stops and warehouse (Yang et al., 2014). These can be used to estimate fuel consumption and emissions.

GPS data analysis has been used to estimate the variations in route characteristics of various categories of freight including couriers and food deliveries in Madrid (Comendador, 2012). Considerable differences in the daily stops, stop durations and distances between stops as well as use of vans (driving time/time stopped) were observed.

GPS data has been utilised to compare the performance of delivery routes occurring at different times of the day to estimate the benefits of off-hours deliveries (Holguín-Veras et al., 2011; 2018; Brenden et al., 2017). GPS data was used to estimate emissions by analysing the dynamic temporal speed patterns of freight vehicles. This has been performed for evaluating the benefits of off-hour deliveries (OHD), where significant reduction of emissions during off-peak periods was estimated based on higher vehicle speeds with less interrupted traffic flow.

An analysis of walking distances as well as vehicle distances travelled by courier drivers in central city areas has been conducted using GPS data (Thompson et al., 2018). This allowed driving and walking routes of a courier in Sydney's CBD, where the walking routes from loading zones to receivers can be distinguished from the driving routes. Maps based on GPS data can be used to analyse walking tours from loading zones.

GPS data allowed the efficiency of courier routes to be determined as well as aid the planning and management of loading zones. Experiments comparing the efficiency of various modes for deliveries within Sydney's CBD were conducted using GPS devices. The routes of cargo bikes, walkers with trollies, and vans were analysed.

GPS has been shown to be an efficient means of collecting data about routes in central city areas (Leonardi & Browne, 2009). More recently GPS has been used to evaluate the performance of IT solutions for couriers (Leonardi & Yamada, 2018).

3.2.3 Road network performance

Freight telematics has recently been used to collate speeds experienced by freight vehicles on individual road segments in Australian cities. GPS traces from a sample of private operator's freight vehicles are used to analyse the performance of truck journeys on a wide range of road network elements (Green and Mitchell, 2018; Austroads, 2019). A similar analysis at a city level has also been conducted in Sao Paulo (Laranjeiro et al., 2019).

Information on how heavy vehicles use the road network and where heavy vehicle congestion occurs assists the government in making decisions relating to road infrastructure spending in terms of where it is best directed, when it is required and the level of investment needed. Carriers can also use these analytics tools to assist in trip planning.

Several measures of performance for routes are generated including median and interquartile range of travel speeds as well as Mean Excess Time Ratio (METR), the mean hourly ratio of interquartile range to the smallest observed interquartile range, and the Mean Excess Uncertainty Ratio (MEUR), the mean hourly ratio of interquartile range to the smallest observed interquartile range. Aggregate performance measures for each state capital city are represented by the mean of METR and MEUR weighted by the distances and volumes of traffic travelling on each route (BITRE, 2023).

Detailed route-specific freight vehicle travel times and congestion measures, including median and interquartile range travel times, for each route in each capital city are produced. The routes are grouped by city and for each route the outputs comprise a route map, table of median and variation in travel times, and graphs showing the hourly distribution of median and interquartile range of travel time.

Maps can be produced to visualise the speeds experienced on links in major metropolitan areas such as Sydney and Melbourne. Profiles of speeds by time of day on specific routes can also be presented.

Maps displaying road segment performance as well as trip durations of user specified routes can be generated from the GPS data of truck movement on major roads. These tools can be used to investigate how congestion impacts freight flows. A number of visualisation tools were developed for conducting route analysis (National Freight Data Hub). Trip duration profiles can also be produced for months and days of the week.

The telematics programme allowed the significant effect of COVID-19 to be visualised (BITRE, 2022). It was found that peaks in freight vehicle average travel times coincide with high commuter volumes. Sydney and Melbourne experienced the largest increases in vehicle congestion between 2021 and 2020. These cities also experienced the most significant COVID-19 lockdowns and the largest falls in congestion between 2019 and 2020.

Increases in congestion in 2021 were identified to highlight the reversal of the impact of COVID-19 related lockdowns in 2020. Reduced congestion from lockdowns in 2020 were shown to be largely reversed, with freight vehicle congestion returning to 2019 levels. Time-of-day profiles of median and interquartile range travel times for routes were produced to allow the effects of congestion to be illustrated.

3.2.4 Freight vehicle demand

GPS data allows information to be generated for analysing urban freight demand patterns. A recent study analysing freight demand patterns was undertaken using GPS data generated from freight vehicles as part of Belgium's dynamic road pricing scheme (Hadavi et al., 2019). Indicators were defined based on the objectives of the strategic freight plan in Brussels. A methodology was developed for analysing the flows that consisted of freight movement understanding, data understanding, data preparation, visualisation, modelling through aggregation and post processing.

Hourly profiles of the total freight demand and distance driven by trucks in the Brussels City Region (BCR) were produced. Histograms illustrating the distribution of kilometres driven, number of stops per vehicle and average distance between stops were generated. GPS data also allowed analysis of origins and destinations within the BCR to be undertaken.

3.3 GEOGRAPHIC INFORMATION SYSTEMS (GIS)

3.3.1 Introduction

Geographic Information Systems (GIS) provide a means of analysing digital spatial data to provide information to decision makers to improve urban freight systems. GIS can be used to produce a range of information relating to:

i. Demand patterns,
ii. Impacts: noise, air quality and crashes,
iii. Freight networks: characteristics and performance,
iv. Output from models,
v. Vehicle routes, and
vi. Land use: type, density and demographics.

GIS can enhance the planning and management of urban freight systems by integrating a range of location specific demand, supply and impact data to provide information to identify problems and evaluate initiatives.

3.3.2 Overlaying

GIS provides a powerful tool for producing maps displaying patterns of characteristics of facilities and areas within metropolitan regions. GIS can provide an effective platform for supporting the collection and collation of data for urban freight transport. Golini et al. (2018) describes a framework for integrating 28 different layers that provide an efficient tool for retrieving and displaying information. This facilitates stakeholder engagement for identifying and understanding problems as well as evaluating solutions. Maps showing the location of demand, traffic networks, warehouses and restricted areas were produced. The framework has been applied in Bergamo and Luxembourg.

An analysis of the suitability of areas within inner Melbourne was undertaken using multi-criteria analysis with the results presented using raster based GIS (Aljohani & Thompson, 2020). Heat maps showing the spatial variation in performance for 11 criteria, including land-use zones, proximity to freight corridors, rental costs and impact on residents were produced. An overall suitability map for candidate sites was also generated. An analysis process comprising six stages was undertaken:

1. Identification of decision criteria for location assessment,
2. Determination of criteria weights (using the Fuzzy Analytical Hierarchical Process),
3. Construction of map layers for each decision criteria for each grid in the study area (using ArcGIS Analytical Tools),
4. Normalisation values of map criteria (using fuzzy logic),
5. Generation of land suitability map to identify candidate sites, and
6. Evaluation of candidate sites (using TOPSIS).

The variation in demand for deliveries within urban areas can be undertaken using heatmaps or using symbols to represent the magnitude of customers in regions within cities. This can provide insights into the likely impacts on distribution patterns when warehouses or distribution centres are relocated. GIS is a powerful tool to understand the transport impacts of logistics sprawl. Maps depicting the demand patterns within Melbourne for fruit and vegetable stores to illustrate the impacts of relocating the wholesale market from West Melbourne to Epping were generated (Aljohani & Thompson, 2018). Retail demand within regions of Melbourne's metropolitan area was depicted using colour to indicate regions and width of circles to show the relative demand levels.

GIS was used to undertake analysis of how COVID-19 changed the demand for B2C goods in Sydney's metropolitan area (Kahalimoghadam et al., 2021). The significant reduction in demand to the CBD as well as considerable growth experienced in many residential areas in the outer regions especially in the south-west can be produced using GIS (Figure 3.1).

Using demand data from a major B2C and a major B2B and B2B carrier, maps were produced by GIS to visualise the changes in distribution patterns. Monthly demand for parcels at the postcode level for 2019 and 2020 were analysed. Nine categories of ranges were used to compare monthly demand levels.

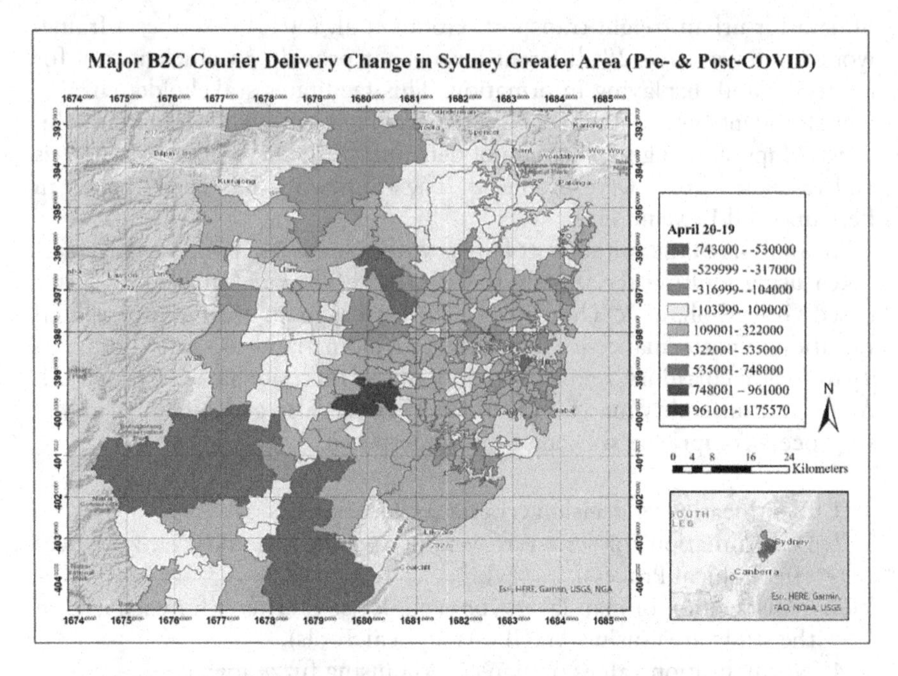

Figure 3.1 Effects of COVID on distribution patterns in Sydney.

GIS was used to conduct spatial analysis of the changes in demand for parcel distribution within the Sydney metropolitan area due to COVID-19. Changes to demand patterns due to lockdowns and increased working from home were able to be observed. The increased demand for home deliveries due to the metropolitan areas and decrease in deliveries to the Central Business District (CBD) was able to be visualised.

GIS provide tools for investigating areas adjoining road networks that can be used to predict the impacts of freight in urban areas. Buffers can be created to identify the location of households in residential areas that are within defined distances from roadways that will experience impacts of freight vehicles such as noise and emission of trucks. Maps of delivery areas from loading zones can also be created to aid delivery planning (Guerlain et al., 2016). GIS allows the relationship between the location of loading zones and freight generators to be identified.

A multi-objective optimisation model incorporating impacts on public health from emissions based on population exposure from the location of freight transport facilities has been developed in Can Tho City, Vietnam (Olapiriyakul & Nguyen, 2019). GIS tools were used to illustrate the potential health impacts of the location of manufacturing plants generated from the model. Using GIS, the affected areas were identified by plotting buffers in the vicinity of transport routes between collection centres, plants and warehouses. Zones with higher densities were also highlighted to investigate health impacts.

Spatial analytics was used to investigate how to improve retail networks in Melbourne (Chhetri et al., 2017). Maps were constructed to highlight which customers are closest to each store as well as which stores are currently servicing customers. Colours were used to highlight the customers serviced by individual stores. Isochrones were used to determine zones based on 15-minute drive times from stores.

GIS allowed the analysis of market areas to be undertaken, illustrating the significant overlap of market areas and highlighting that some stores compete for business. Improved understanding of the spatial relationship between stores and customers was gained. This allowed new sets of market catchments for distribution to be investigated to better cater for customer service requirements.

3.3.3 HazMat vehicle route planning

Trucks transporting hazardous material (hazmat) in urban areas present a major health risk for residents. There is a need to define routes to minimise exposure to the population. A spatial decision support system (SDSS) for undertaking quantitative risk analysis and calculating minimum-risk paths as well as visualising these on digital maps has been developed (Ak et al., 2020). GIS was used to create risk maps for routing scenarios to improve the planning of hazmat transportation in Istanbul. Optimal traffic paths based on risk criteria such as time-based risk, population exposure, societal risk and incident probability as well as the shortest and fastest paths between specific origin and destinations can be calculated.

GIS has been used to interactively display vehicle routes in urban networks for various costs such as vehicle capacity and operating costs as well as environmental costs (Gaudron et al., 2020). A GIS was used to display the roads, locations of clients and warehouse as well as vehicle routes based on a set of adjustable costs and van parameters. Operational parameters such as fuel costs and drop-off duration can be varied. A number of van characteristics such as fuel consumption and emission rates, capacity, speed as well as climate change and public health costs can be adjusted. Plots of the van routes for specific sets of parameters can highlight the potential improvements from introducing new types of vehicles in urban distribution.

Animation of freight vehicle movements in cities from agent based models developed in AnyLogic can be displayed using digital maps (Raoui, 2018; Gusah et al., 2019). This can aid verification and validation of simulation models.

An interactive mapping tool has been created to assist carriers with planning vehicle routes based on the carriers vehicle classification (NVHR, 2018). This system allows journeys to be created for user specified origins and destinations to identify the approved network and determine where an application is required to be submitted (Figure 3.2).

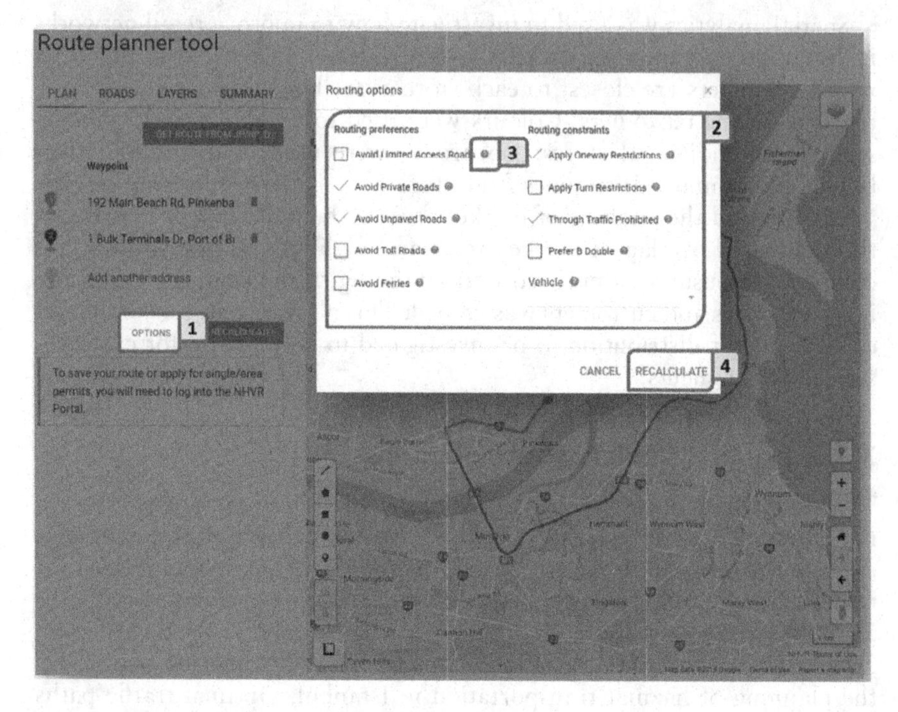

Figure 3.2 Route planning.

GIS was used to evaluate the feasibility of crowd shipping using public transport and active transport for University students considering the proximity of delivery points and home addresses and students' travel patterns (Giuffrida et al., 2021). Threshold areas were computed on the real pedestrian network based on OpenStreetMap (OSM) data, by creating iso-distance areas from the origins using the ORS Tools plugin in the QGIS environment.

GIS provides a useful tool for investigating logistics sprawl (Aljohani & Thompson, 2016). The number of warehouses and establishments per zip code including the barycenter, or geographic mean (distance weighted), were calculated and plotted for both Los Angeles and Seattle in 1998 and 2009 using ArcGIS procedures (Dablanc et al., 2014). It was noted that the average distance of warehousing establishments from their barycenter increased by 6 miles during that period.

3.4 DATA AND SOFTWARE

QGIS is a popular open source GIS that has recently been used in a number of urban freight studies (Teoh et al., 2018; Ewbank et al., 2020; Carrese et al., 2021, de Oliveira, et al., 2021; Pirra et al., 2021). ArcGIS has also been widely used to investigate urban freight systems, including crashes (McDonald et al., 2018), city allies (Machado-León et al., 2020) and loading zones (Alho et al., 2018b).

OpenStreetMap provides a rich source of transport network data. Road network data for intersections and street segments in GIS format for many cities throughout can be obtained from OpenStreetMap (OSM).

Many governments have implemented OpenData policies and have made available geographically coded datasets that can be useful in undertaking urban freight studies. Data relating to urban freight networks, volumes, crashes and travel times are able to downloaded for many major cities.

3.5 CONCLUSIONS

Decision makers require advanced spatial analytical tools for understanding urban freight systems as well as evaluating potential solutions. GPS provides a rich source of data regarding the movement of freight vehicles that can assist planners and managers of urban freight systems identify problems and better understand vehicle movement patterns.

GIS allows a vast range of network, demand and impact data to be integrated and mapped to consider multiple stakeholders interests that can aid public sector policy formulation and evaluation. High-quality maps can be produced to aid the identification of spatial relationships.

Recent developments in software systems allow more interactive analysis as well as more dynamic data to be visualised that can improve the range of information available for designing and managing urban freight systems.

REFERENCES

Ak, R., Bahrami, M. and Bozkaya, B. (2020). A time-based model and GIS framework for assessing hazardous materials transportation risk in urban areas, *Journal of Transport & Health*, **19**, 100943.

Alho, R.A., You, L., Lu F., Cheah L., Zhao F. and Ben-Akiva, M. (2018a). Next-generation freight vehicle surveys: Supplementing truck GPS tracking with a driver activity survey, In: *21st International Conference on Intelligent Transportation Systems (ITSC), Maui, Hawaii.*

Alho, A., de Abreu e Silva, J., de Sousa, J. and Blanco, E. (2018b). Improving mobility by optimizing the number, location and usage of loading/unloading bays for urban freight vehicles, *Transportation Research Part D*, **61**, 3–18.

Aljohani, K. and Thompson, R.G. (2016). Impacts of logistics sprawl on the urban environment and logistics: Taxonomy and review of Literature, *Journal of Transport Geography*, **57**, 255–263.

Aljohani, K. and Thompson, R.G. (2018). Impacts of relocating a logistics facility on last food miles – The case of Melbourne's fruit & vegetable wholesale market, *Case Studies on Transport Policy*, **6**, 279–288.

Aljohani, K. and Thompson, R.G. (2020). A multi-criteria spatial evaluation framework to optimise the siting of freight consolidation facilities in inner-city areas, *Transportation Research Part A*, **138**, 51–69.

Austroads (2019). *Key Freight Routes - Heavy Vehicle Usage Data Project*, AP-R602–19, Austroads, Sydney.

BITRE (2020). *Freight route performance under COVID-19*, Information Sheet 107, Bureau of Infrastructure, Transport and Regional Economics (BITRE), Australian Government, Canberra. freight-route-performance.

BITRE (2023). *Road Freight Congestion Report 2022*, Bureau of Infrastructure, Transport and Regional Economics (BITRE), Australian Government, Canberra, Congestion-report-2022.

Brenden, A.P., Koutoulas, A., Fu, J., Rumpler, R., Sanchez-Diaz, I., Behrends, S., Glav, R., Cederstav, F. and Brolinsson, M. (2017). *Off-peak City Logistics - A Case Study in Stockholm*, KTH Royal Institute of Technology.

Carrese, F., Colombaroni, C. and Fusco, G. (2021). Accessibility analysis for urban freight transport with electric vehicles, *Transportation Research Procedia*, **52**, 3–10.

Chhetri, P., Kam, B., Lau, K. H., Corbitt, B. and Cheong, F. (2017). Improving service responsiveness and delivery efficiency of retail networks, A case study of Melbourne, *International Journal of Retail & Distribution Management*, **45**, 3, 271–291.

Comendador, J., López-Lambas, M.E. and Monzón, A. (2012). A GPS analysis for urban freight distribution, *Procedia - Social and Behavioral Sciences*, **39**, 521–533.

Dablanc, L., Ogilvie, S. and Goodchild, A. (2014). Logistics sprawl differential warehousing development patterns in Los Angeles, California, and Seattle, Washington, *Transportation Research Record*, **2410**, 105–112.

de Oliveira, I., de Oliveira, L. and de Albuquerque No´brega, R. (2021). Applying the maximum entropy model to urban freight transportation planning: An exploratory analysis of warehouse location in the Belo Horizonte Metropolitan Region, *Transportation Research Record*, **2675**, 65–79.

Ewbank, H., Vieira, J., Fransoo, J. and Ferreira, M. (2020). The impact of urban freight transport and mobility on transport externalities in the SPMR, *Transportation Research Procedia*, **46**, 101–108.

Fransoo, J.C., M.G. Cedillo-Campos and K.M. Gamez-Perez (2019). Dedicated unloading bays by field experimentation, *Proceedings 11th International Conference on City Logistics*, Dubrovnik, Institute for City Logistics.

Gaudron, A., Tamayo, S., de La Fortelle, A. (2020). Interactive simulation for collective decision making in city logistics, *Transportation Research Procedia*, **46**, 157spor.

Giuffrida, N., Le Pira, M., Fazio, M., Inturri, G. and Ignaccolo, M. (2021). On the spatial feasibility of crowdshipping services in university communities, *Transportation Research Procedia*, **52**, 19–26.

Golini, R., Guerlain, C., Lagorio, A. and Pinto, R. (2018). An assessment framework to support collective decision making on urban freight transport, *Transport*, **33**, 4, 890–901.

Green, R. and Mitchell, D. (2018). Adapting truck GPS data for freight metrics, *Australian Transport Research Forum*, Darwin, ATRF 2018, Paper 18.

Guerlain, C., Cortina, S. and Renault, S. (2016). Towards a collaborative geographical information system to support collective decision making for urban logistics initiative, *Transportation Research Procedia*, **12**, 634–643.

Gusah, L., Cameron-Rogers, R. and Thompson, R.G. (2019). A systems analysis of empty container logistics - A case study of Melbourne Australia, Green Logistics for Greener Cities, Green Cities 2018, 13–14th September, Szczecin, *Transportation Research Procedia*, **39**, 92–103.

Hadavi, S., Verlinde, S., Verbeke, W., Macharis, C. and Guns, T. (2019). Monitoring urban-freight transport based on GPS trajectories of heavy-goods vehicles, *IEEE Transactions on Intelligent Transport Systems*, **20**, 10, 3747–3758.

Holguín-Veras, J., Ozbay, K., Kornhauser, A., Brom, M.A., Iyer, S., Yushimito, W.F., Ukkusuri, S., Allen, B. and Silas, M.A. (2011). Overall impacts of off-hour delivery programs in New York City Metropolitan Area, *Transportation Research Record*, **2238**, 68–76.

Holguín-Veras, J., Encarnación, T., González-Calderón, C.A., Winebrake, J., Wang, C., Kyle, S., Herazo-Padilla, N., Kalahasthi, L., Adarme, W., Cantillo, V., Yoshizaki, H. and Garrido, R. (2018). Direct impacts of off-hour deliveries on urban freight emissions, *Transportation Research Part D*, **61**, 84–103.

Kahalimoghadam, M., Stokoe, M., Thompson, R.G. and Rajabifard, A. (2021). The impact of COVID-19 pandemic on parcel delivery pattern in Sydney, In: *Proceedings, Australian Transport Research Forum (ATRF)*, Brisbane, December 8–10, 2021.

Laranjeiro, P.F., D. Merchan, L.A. Godoy, M. Giannotti, H.T.Y. Yoshizaki, M. Winkenbach and C.B. Cunha (2019). Using GPS data to explore speed patterns and temporal fluctuations in urban logistics: The case of Sao Paulo, Brazil, *Journal of Transport Geography*, **76**, 114–129.

Leonardi, J. and Yamada, T. (2018). Can routing systems surpass the routing knowledge of an experienced driver in urban deliveries? In: Taniguchi, E. and Thompson, R.G. (eds.), *City Logistics I- New Opportunities and Challenges*, ISTE, 381–400, Wiley.

Leonardi, J. and Browne, M. (2009). *Review and Test of GPS Use in Urban Freight Data Collection and Analysis*. London Data and Knowledge Centre, Transport for London.

Machado-León, J. L., del Carmen Girón-Valderrama, G. and Goodchild, A. (2020). Bringing alleys to light: An urban freight infrastructure viewpoint, *Cities*, **105**, 102847.

McDonald, N., Yuan, Q. and Naumann, R. (2019). Urban freight and road safety in the era of e-commerce, *Traffic Injury Prevention*, **20**, 7, 764–770.

NVHR (2018). NHVR Route Planner Tool User Guide, Version 2, National Heavy Vehicle Regulator, Brisbane.

Olapiriyakul, S. and Nguyen, T. T. (2019). Land use and public health impact assessment in a supply chain network design problem: A case study, *Journal of Transport Geography*, **75**, 70–81.

OSM. OpenStreetMap, www.openstreetmap.org. (accessed 24th May 2023).

Pirra, M., Carboni, A. and Deflorio, F. (2021). Freight delivery services in urban areas: Monitoring accessibility from vehicle traces and road network modelling, *Research in Transportation Business & Management* (In Press).

Raoui, H. E. L, Oudani, M. and Alaoui, A. EL H. (2018). ABM-GIS simulation for urban freight distribution of perishable food, *MATEC Web of Conferences 200, IWTSCE'18*.

Sanchez-Diaz, I., Palacios-Arguello, L., Levandi, A., Mardberg, J., Basso, R. and Behrends, S. (2019). A time-efficiency study of medium duty trucks delivering in urban environments, *Proceedings 11th International Conference on City Logistics*, Dubrovnik, Institute for City Logistics.

Siripirote, T., Sumalee, A. and Ho, H.W. (2020). Statistical estimation of freight activity analytics from Global Positioning System data of trucks, *Transportation Research Part E*, **140**, 101986.

Teoh, T., Kunze, O., Teo, C. and Wong, Y. D. (2018). Decarbonisation of urban freight transport using electric vehicles and opportunity charging, *Sustainability*, **10**, 3258.

Thompson, R.G., Zhang, L. and Stokoe, M. (2018). Optimising courier routes in central business districts, In: Taniguchi, E. and Thompson, R.G. (eds.), *City Logistics I- New Opportunities and Challenges*, ISTE, Wiley, 325–342.

Toilier, F. and Gradrat, M. (2019). Driver surveys vs vehicle GPS data: Comparability, strengths and weaknesses of the two sources in order to characterize the drivers' journeys, *Proceedings 11th International Conference on City Logistics*, Dubrovnik, Institute for City Logistics.

Yang, X., Sun, Z., Ban, X. J., Wojtowicz, J. and Holguín-Veras, J. (2014). Urban freight performance evaluation using GPS data, *Proceedings* 93rd Annual Meeting Transportation Research Board (TRB), **600**, 45.

Chapter 4

Optimisation

4.1 INTRODUCTION

Logistics has a considerable share in the end-user cost of products and services, therefore, it impacts both consumers and suppliers. Ever increasing competition in consumer and service industry requires the concerned stakeholders to optimise the logistics cost to maintain sufficient profits. There exists a vast body of literature focusing on optimisation in all aspects of logistics, such as the location of facilities (*e.g.*, factories, warehouses, and depots), inventory management, and the transportation of raw materials and finished goods. In this chapter, we introduce some of the key optimisation problems in city logistics and their solution algorithms.

4.2 VEHICLE ROUTING PROBLEM

The vehicle routing problem (VRP) appears naturally in the last mile of urban deliveries, where customers demand less than truck load (LTL) quantities to be delivered or picked up. The alternative method, *i.e.*, dedicated one-to-one deliveries would be very expensive. Therefore, merging one-to-one routes together without violating the capacity of the delivery vehicle will results in savings. This idea is used in one of the earlier solution algorithms of the VRP, *i.e.*, the saving's algorithm (Clark & Wright, 1964). The VRP consists of finding a set of minimum cost routes for k vehicles stationed at central depot, to cover all demands d_i of geographically located n customers with a constraint that the sum of demands along a route shall be less than the vehicle capacity q. Since its inception in 1959 (Dantzig & Ramser, 1960), the VRP has seen inclusion of various practical constraints such as time windows (VRPTW) (Solomon, 1987), consideration of electric vehicles (EVRPTW) (Conrad & Figliozzi, 2011) and integration of delivery drones and robots (Deng et al., 2020).

DOI: 10.1201/9781003261513-5

4.2.1 The vehicle routing problem with time windows

The Vehicle Routing and Scheduling Problem with Time Windows (VRPTW) arises due to the addition of the *time windows constraint* to the VRP definition. This constraint ensures the start of service at each customer i within its pre-specified time window $[a_i, b_i]$, where a_i specifies the earliest possible service time and b_i as the desired latest possible service time at the concerned customer location. Figure 4.1 shows a typical VRPTW instance and a possible solution.

The VRPTW is defined on a directed graph $G=(V, A)$. The vertex set V includes the depot vertex 0 and a set of customers $C=\{1, 2,..., n\}$. The set of identical vehicles with capacity q, stationed at the depot is represented by K. With every vertex of V there is associated a demand d_i, with $d_0=0$, and a time window $[a_i, b_i]$ representing the earliest and the latest possible service start times. A travel cost $c_{ij,}$ as well as a travel time $t_{ij,}$ are associated with each arc of the graph. Sometimes the quantity t_{ij} includes the travel time on arc (i, j) and the service time at vertex i. The arc set A consists of all feasible arcs (i, j) satisfying the inequality $a_i+t_{ij}\leq b_i$, $i, j \in V$. The service start time at a vertex $j \in C$ by a vehicle $k \in K$, is defined by s_{jk}.

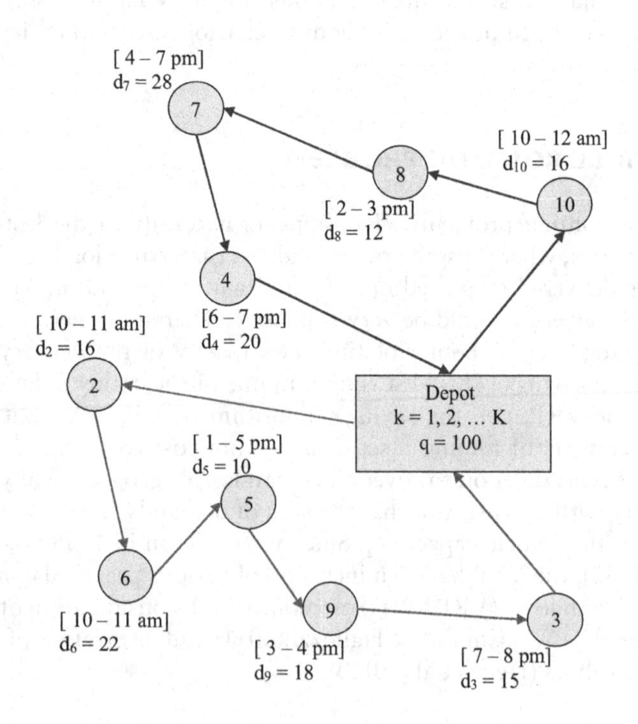

Figure 4.1 A typical VRPTW solution.

Using the definitions mentioned in the previous paragraph, the VRPTW can be described by a three-index mathematical formulation as follows:

$$\min \sum_{k \in K} \sum_{(i,j) \in A} c_{ij} x_{ijk} \tag{4.1}$$

subject to

$$\sum_{k \in K} \sum_{j \in V} x_{ijk} = 1 \qquad \forall\, i \in C \tag{4.2}$$

$$\sum_{i \in C} d_i \sum_{j \in V} x_{ijk} \leq q \qquad \forall\, k \in K \tag{4.3}$$

$$\sum_{j \in V} x_{0jk} = 1 \qquad \forall\, k \in K \tag{4.4}$$

$$\sum_{i \in V} x_{ihk} - \sum_{j \in V} x_{hjk} = 0 \qquad \forall\, h \in C, \qquad \forall\, k \in K \tag{4.5}$$

$$\sum_{i \in V} x_{i0k} = 1 \qquad \forall\, k \in K \tag{4.6}$$

$$s_{ik} + t_{ij} - s_{jk} \leq \left(1 - x_{ijk}\right) M_{ijk} \qquad \forall\, (i,j) \in A, \qquad \forall\, k \in K \tag{4.7}$$

$$a_i \leq s_{ik} \leq b_i \qquad \forall\, i \in V, \qquad \forall\, k \in K \tag{4.8}$$

$$x_{ijk} \in \{0,\ 1\} \qquad \forall\, (i,j) \in A, \qquad \forall\, k \in K \tag{4.9}$$

The model contains two decision variables, s_{jk} the service start time at node j, if j belongs to the route serviced by the vehicle k. The second decision variable is x_{ijk}, which determines whether arc (i, j) is used in the solution (x_{ijk}=1) or not (x_{ijk}=0). M_{ijk} is a large constant. The objective function (Eq. 4.1) minimises the total cost of the solution. The assignment constraint (Eq. 4.2) ensures that every customer must only be serviced once while the capacity constraint (Eq. 4.3) keeps the total load along a route within the vehicle capacity q. Constraints given by (Eqs. 4.4–4.6) are flow conservation constraints, stating that every vehicle k must start and end at the depot while along the route if it *travels to* (inwards) any customer h it must also *travel from* customer h (outwards). The time windows constraints (Eqs. 4.7 and 4.8) specify that if a vehicle travels from i to j, service at j cannot start earlier than that at i, and restricts the service start time at all vertices within their specified time windows $[a_i, b_i]$. The time windows constraints also ensures that there are no subtours in the feasible solution. Finally, the integrality constraint is represented by (Eq. 4.9).

The nature and different treatments of the user defined time windows produce different variants of the VRPTW. With the Vehicle Routing and Scheduling Problem with Hard Time Windows (VRPHTW), delivery of goods outside the time windows is not allowed at all. A more practical variant of the VRPTW is the Vehicle Routing and Scheduling Problem with Soft Time Windows (VRPSTW), where deliveries are still possible outside the time windows with some penalty costs. The cost (c'_{ij}) and the service start time (s_{jk}) are calculated based on the arrival time of vehicle at customer j (Bhusiri et al., 2014). Sometimes the early arrival penalty is dropped (allowing waiting at no cost) to keep the cost as monotonically increasing with time obeying the triangular inequality ($c_{ij} \leq c_{ih} + c_{hj}$) (Qureshi et al., 2009a).

4.2.2 Dynamic VRPs

Many inputs in the classical VRP can change during the operation day (or the scheduling horizon) such as number of the customers due to new incoming requests, service times at customers due to their availability, and travel times due to unexpected incidents. The dynamic vehicle routing reacts to such a change in information and re-optimises the system (Psaraftis, 1995). The scheduling horizon is divided into many sections either at a priori fixed time durations (Chen & Xu, 2006) or when some new information is received (such as new customer requests or service completion at a customer by a delivery vehicle). Usually, each section of the scheduling horizon is treated as a static VRP instance consisting of the available information of customers, vehicle resources, and the network. A new section unfolds at the end of the previous one in a rolling horizon fashion. Sometimes predefined information or a guess is available for the changing information environment, in such cases dynamic models often employ strategies to get benefit by considering it during the planning. For example, the pre-positioning of idle vehicles in areas of high expectancy of new requests (Branchini et al., 2009). Larsen et al. (2002) presented a classification of dynamic routing systems based on the idea of available reaction time, which is the difference between the times when a change in the information is first available to the time a system must respond to it.

The vehicle routing problem with dynamic travel time considers the real-time traffic data from the city-wide intelligent transportation systems (ITS). Generally, data from these ITS systems is very large and poses difficulties in its utilisation, such as data cleansing and abstracting; Fleischmann et al. (2004) used such travel time data from a system implemented in Berlin, Germany and discussed the related issues. Taniguchi and Shimamoto

(2004) have used data from an ITS system in Japan and proposed a genetic algorithm to solve the related VRPTW with dynamic travel times. Qureshi et al. (2012) used micro-simulation software to generate dynamic travel time data under various incident scenarios.

4.2.3 Stochastic VRPs

In contrast to the dynamic VRPs, where the input data (such as customers' demand or travel time) keeps on changing unexpectedly, Stochastic Vehicle Routing Problem (SVRP) assumes that a probability distribution is available for these inputs. Therefore, while the actual values of such inputs are revealed at the time of service, an initial guess (in form of probability distribution) is available at the planning stage. Stochasticity in travel times can occur due to varied traffic conditions in different part of a day (such as morning or evening peeks). Similarly, delivery demand can be based on some historical data in case of regular, repetitive routing systems such as deliveries of food items, milk or other beverages, pickup of cash from banks at the end of the day, and in heating oil's distribution to households.

Dror and Trudeau (1986), Christiansen and Lysgaard (2007) and Tan et al. (2007) presented the stochastic customers' demand VRPs, where each customer demand is represented by an independent random variable; whereas, Stewart and Golden (1983) solved a case of correlated stochastic demands. In the Vehicle Routing Problem with Stochastic Customers and Demand (SVRP-CD), the presence of the customer itself is given with a probability value along with the probability distribution of its demand (Gendreau et al., 1995, 1996a). The presence of a customer is revealed when the service at the predecessor node begins while its demand is made available when a vehicle reaches it. Gendreau et al. (1996b) provided a comprehensive and concise list of various distributions used for modelling the stochastic demand such as normal, Poisson and binomial.

As the probability distribution is available for the stochastic inputs, "a priori solution" (Bertsimas et al., 1990) is usually obtained at the start of the scheduling horizon. It is then simulated by sampling the probability distribution (termed as a "state of world" (Taniguchi et al., 2001)) representing a possible actual value of the stochastic variable. For example, in the case of the stochastic demand, the sampled value represents the actual demand of the customers, and it can be larger or smaller than its expected demand (considered in the a priori solution earlier). If the sampled value turns out to be larger, it can cause a "route failure", *i.e.*, the current residual vehicle capacity is insufficient to meet this demand. In the chance constrained modelling, the probability of such route failure is restricted to be within a pre-specified limit (say β). A recourse action (or recourse policy) is used to repair the route failure, such as the vehicle returning to the depot, unloading its content to regain the capacity and then continuing with either the same a priori route (static problem) or with a different route (dynamic problem). The recourse modelling of the SVRP assigns a penalty cost to this recourse and incorporates it into the objective function. Uncertainties in customer demand have received more attention and the literature considering the Vehicle Routing Problem with Stochastic Travel Times Demand (SVRP-T) is rather scant (Laporte et al., 1992).

4.2.4 Green VRPs

There has been considerable research interest in incorporating the environmental impact of urban freight in the vehicle routing model. For example, Bektas and Leporte (2011) formulated a pollution routing problem (PRP) considering the CO_2 emissions based on the fuel consumption along the arcs depending on the vehicle speed and vehicle load. Works of Kuo and Wang (2011), Jabali et al. (2012), Xiao et al. (2012) and Kramer et al. (2015) can be included in the same class, although they used somewhat different equations to calculate the fuel consumption and corresponding CO_2 emissions. Afshar-Bakeshloo et al. (2016) added a third objective of customers' satisfaction along with distance and emissions (fuel consumption). Lin et al. (2014) provided an exhaustive review of the GVRP and other variants of the VRP.

Electric vehicles (EVs), with their low emissions and operating costs have been thought as an alternative to conventional vehicles to reduce the environmental footprint of VRPs. However, their relatively high initial cost (Vermie, 2002), limited driving range, speed and acceleration make them less attractive for the freight carriers (Jeeninga et al., 2002). Many comparisons of EVs with conventional vehicles have been presented in the literature. Davis and Figliozzi (2013) compared three different types of electric vehicles with conventional trucks under varying scenarios. They showed that electric trucks can be a viable alternative in the presence of the following conditions: Maximum distance travelled matches the range of the EVs (taken as 100 miles, about 160 km), low speed or congestion, and frequent stops. A similar conclusion was reported by Lee et al. (2013), who compared electric vehicles and conventional diesel trucks in terms of energy consumption, greenhouse gas emissions and ownership costs. Macharis et al. (2013), analysed the competitiveness of electric vehicles from an economic point of view considering subsidies, taxes, insurance, maintenance, and car inspection costs along with initial cost and fuel consumption in the Brussels-Capital Region, and found that EVs are advantageous in the lower payload range (less than 1,000 kg). However, based on the life-cost analysis, Giordano et al. (2018) concluded that appropriate policies are still needed to offset the cost differences to replace old diesel vans with EVs. In their paper they mention that in some cities (such as Oslo and London), electric vans are already cheaper than diesel vans on an equivalent annual costs basis due to policies and incentives already implemented, but in other countries appropriate push measures (taxation and regulations on diesel trucks) and pull measures (subsidies and de-regulations on EVs) are still required. Lebeau et al. (2016) presented a conjoint-based choice model considering the daily range, charging time, environmental performance, type of vehicle, purchase cost and operating cost as the attributes defining the vehicle choice. Sen et al. (2017) brought more precision to this comparative analysis by comparing different types of electric trucks (with 270 kWh motors and 400 kWh motors) with not only conventional heavy-duty trucks but also included hybrid, CNG, biodiesel trucks. Their study concluded that if the electricity generation-mix is based on sustainable/renewable sources, electric heavy trucks have an advantage over all other types of trucks tested in their study, despite the high initial cost of electric trucks.

An important factor that affects the modelling of EVs is the consideration of battery usage during their operation. Goeke and Schneider (2015) considered a realistic energy consumption model based on the speed, gradient and cargo load distribution; whereas Lebeau et al. (2015) estimated a non-linear regression equation for the energy consumption using experimental results and theoretical values defined by speed, gradient and air drag. Fiori and Marzano (2018) used a trip-based Tank-to-Wheel (TTW) model based on the length of the driving cycle, speed, acceleration, road gradients, and payloads along with vehicle's powertrain characteristics. Basso et al. (2019) first determined the paths between customers based on energy estimation using an approximated driving cycle considering speed, topography, powertrain efficiency and the effect of acceleration and braking at traffic lights and intersections. The order of the customers was then optimised in the second stage using energy cost coefficients of these paths along with a linearised function of the weight of the vehicle.

Availability of re-charging possibilities reduces the impact of limited range disadvantage of EVs. Afroditi et al. (2014) presented a mathematical formulation of the electric vehicle routing problem with time windows (EVRPTW) considering a predefined range and possible recharges along their routes at charging stations. Conrad and Figliozzi (2011) presented an EVRPTW, where vehicles can recharge at customer locations. Montoya et al. (2017) considered the nonlinear behaviour of the charging process that depends on state of charge (SOC), voltage and current relationship using a piecewise linear approximation. Keskin et al. (2019) considered time-dependent charging times based on the assumption that EVs might need to wait in a queue at public charging stations.

4.2.5 Heterogeneous vehicle routing problem

The Heterogeneous fleet Vehicle Routing Problem (HVRP) deals with the availability of vehicles of different capacities at a central depot (Choi & Tcha, 2007). Different vehicles in the HVRP can be classified depending on their load carrying capacities, types of products they can carry and their driving range. Usually, the fixed costs are different for each category. The travel cost matrices can also be different based on their ability to use the complete or partial road network or the links connecting customers and depots. With the emergence of new vehicle technologies such as electric vehicles, (electric/manual) cargo bikes, drones and robots, the HVRP has become more prominent in the research as the routing systems using combinations of these vehicles can become operational in the near future. Most of the routing systems integrating drones/autonomous delivery robots (ADR) presented in Chapter 8 can be categorised as HVRPs. For example, Deng et al. (2020) examined a combination of a heavy capacity resources with unlimited range and a light capacity resource with a limited range (drones, robots and walker/human). Based on their assumptions considering capital cost, capacity and range, they concluded that the human walker/helper can save more costs than both ADRs and drones, whereas ADRs performs better as compared to drones.

In HVRPs, the choice of vehicle type is based on the operational characteristics of the freight carriers, which led to the introduction of various fleet size and fleet mix models based on the vehicle routing framework. For example, van Duin et al. (2013) presented a Fleet Size and Mix Vehicle Routing Problem with Time Windows (FSMVRPTW) considering electric vehicles as one of the vehicle options. Based on a case study in Amsterdam, they concluded that the use of electric vehicles in combination with urban consolidation centres can result in a 19% reduction of vehicle kilometres and 90% reduction in CO_2 emissions. In order to determine an optimal fleet mix and size, Ahani et al. (2016) presented a portfolio optimisation approach considering the total cost and the associated variances of the uncertain parameters such as the initial cost of EVs and the price fluctuation of fossil fuels.

4.3 LOCATION-ROUTING PROBLEM

For a given set of customers, finding the best location of a single (multiple) depot(s) is called the facility location problem (FLP). There is a strong relationship among the location of facilities, inventory levels and transportation decisions in logistics. Better transportation planning alone may not result in an overall cost-efficient logistics system (Ghiani et al., 2013). Therefore, designing an optimum distribution network is essential for a sustainable logistics system. Typical FLPs only consider the direct distances (such as Euclidean or on the network) between the facility to be located and the customers. Traditionally, the FLP is considered a long-term strategic decision whereas, vehicle routing is considered an operational stage issue. Research, however, suggests that solving them together, early in the planning horizon provides a win-win solution from the point of view of public and private costs (Salhi & Nagy, 1999). The combination of the FLP and the VRP is termed the Location-Routing Problem (LRP).

In the beginning, un-capacitated LRPs were introduced, where the depot is assumed to have unlimited capacity and therefore, a single depot is to be located (Laporte & Nobert, 1981). Later, a host of LRPs with multiple capacitated facilities were proposed, including stochastic and dynamic demand sets (Laporte & Dejax, 1989, Laporte et al., 1989). The LRP tries to optimise the decisions for the location of depot(s), assignment of customer nodes to depot(s) and finally the routing of vehicles starting from the selected depots serving the assigned customers. Zografos and Samara (1989) and ReVelle et al. (1991) considered the un-capacitated and capacitated LRPs, respectively, for hazardous waste transportation and disposal. They used multiple objectives, which included disposal risk, routing risk, and travel time. Akca et al. (2009) and Belengure et al. (2011) presented exact solution methods for the capacitated version of the LRP based on branch and price (column generation), and branch and cut methods, respectively.

The location-routing problem with time windows considers the additional challenge of matching the service to customers within their specified time windows with vehicle routes of the classical LRP. Jacobsen and Madsen (1978) studied a newspaper delivery system which considered the time windows and later Srivastava and Benton (1990) studied a LRPTW for general freight distribution. Nikbakhsh & Zegordi (2010) solved the LRP with soft time windows using local search heuristics. Ponboon et al. (2016) presented a branch and price algorithm for the LRPTW. Hof et al. (2017) solved a battery swap station location-routing problem with capacitated electric vehicles (BSS-EV-LRP) and proposed an Adaptive Variable Neighbourhood Search (AVNS) algorithm. While searching the best routes for EVs, the algorithm simultaneously determines the best locations for BSS satisfying the capacity constraint and range constraint of the EVs. Dukkanci et al. (2019) introduced the Green Location-Routing Problem (GLRP), minimising the objective function based on the fixed cost of operating depots, fuel cost and CO_2 emissions while maintaining feasibility with respect to capacity and time windows constraints. Koc (2019) solved a similar GLRP with time windows additionally considering driver costs. Schiffer et al. (2018) introduced a location-routing problem with intermediate facilities with multiple resources (LRPIF-MR). The intermediate facilities were used to provide replenishment for either energy (recharging facilities) or for freight (replenishing facilities) or both (combined facilities). If only the energy resource is considered their problem becomes the same as the BSS-EV-LRP.

4.4 MULTI-ECHELON ROUTING AND LOCATION-ROUTING MODELS

Multi-echelon vehicle routing problems consider classical VRPs at the lowest echelon and create higher echelons by considering the route starting points of the lower echelons. Therefore, in a 2E-VRP, lower echelon routes serving the customers depart from satellites (depots of the classical VRPs), whereas the upper echelon routes connect these satellites *via* routes (possibly using larger vehicles) departing from an upper echelon depot (let's say distribution centre). If a subset of the satellites is to be chosen from a given set of satellite locations, the lower echelon of the 2E-VRP becomes similar to the LRP, however, in some cases these satellites do not have any fixed costs for their establishment (Gonzalez-Feliu, 2013). In a three-echelon system, distribution centres are served (*via* routes) starting from the distribution facility of the third echelon. The system can be generalised to m-echelon systems (Gonzalez-Feliu, 2008). Crainic et al. (2008) presented heuristics-based solutions for the 2E-VRP by dividing the problem at the two levels and solving them sequentially. Consideration of the time windows at customers and/or satellites further complicates decision making for the 2E-VRP (see, *e.g.*, Li et al., 2016) and requires time synchronisation (Jia et al., 2023). Usually, large capacity vehicles are considered at the upper echelon, and it is assumed that they only serve satellites but in some 2E-VRPs customers can also be served by these vehicles, especially if they are not located in the city centre (Anderluh et al. 2021).

A classical LRP lies at the lowest level of the multi-echelon location-routing problem, finding the best routes serving the customers starting from ideally located satellites. These satellites in turn become customers of the second echelon (in 2E-LRP), which finds best locations of distribution centres and links them to satellites with the best routes. Similarly, higher order echelons (mE-LRP) are constructed by finding the best (location of) distribution facilities for the route starting points of the lower levels and the corresponding connecting routes. For example, Lin and Lei (2009) presented a 2E-LRP that integrated decision making for the number and location of the satellites, on routing from a distribution centre to these satellites, and on the routing from each satellite to the customers assigned to it. They also included the possibility of serving customers directly from the distribution centre. Additional decision-making complexity can be added by considering time windows and other VRP-related add-ons. For example, Wang et al. (2018) considered a 2E-LRP with time windows and uncertain demand from customers.

Breunig et al. (2019) introduced the electric two-echelon vehicle routing problem (E2E-VRP). They considered EVs in the lower echelon whereas conventional vehicles were used in the upper echelon to transport commodities from a depot to satellites. They proposed a large neighbourhood search-based metaheuristic to solve the E2E-VRP and an exact mathematical programming algorithm, which uses a decomposition technique and problem-tailored pricing algorithms. Jie et al. (2019), considered large and smaller EVs in both the upper and lower echelons, respectively, with different driving ranges. The EVs of both echelons can visit battery swapping stations (BSS) for recharging. Coelho et al. (2017) introduced the multi-objective Green UAV Routing Problem (GUAVRP) that considered a heterogeneous fleet of Unmanned Aerial Vehicles (UAV). The problems were constructed in a two-echelon fashion, with larger, faster drones with higher capacity working as line haul and smaller drones doing the final deliveries and pickups. Both upper echelon and lower echelon drones can visit charging stations once their battery levels are low.

4.5 HUB LOCATION-ROUTING PROBLEM

The hub location-routing problem (HLRP) is an extension of the classical hub-and-spoke network design problem where the latter is mainly used in situations where goods flow between non-hub nodes (customers) consists of full (or near full) truck loads (Campbell & O'Kelly, 2012). If the flow demanded between non-hub nodes (such as branch offices of intra-city express mail/parcel delivery systems) is mainly in less than truck load (LTL) quantities, it is better to replace the spokes (one-to-one connections between hubs and non-hub nodes) with vehicle routes resulting in the HLRP (Bostel et al., 2015; Çetiner et al., 2010). The HLRP can be efficiently used in warmed-up type systems, where, for example, vehicles depart from hubs and travel across the non-hub nodes while delivering parcels and mail collected in the previous period and collecting parcels (simultaneously or separately) and mail to be delivered in the next period (*e.g.*, next day) (such as in Karimi, 2018). The HLRP thus, integrates the decision making for the location of hubs, the allocation of non-hub nodes (to hubs), the routing of flows between each origin and destination (non-hub nodes), and the visit sequence of non-hub nodes in local tours. It differs from the classical LRP in the sense that LRP doesn't allow goods to flow between depots (or hubs). Usually, non-hub nodes are allocated to a single hub but recently Wu et al. (2022), presented a HLRP model where non-hub nodes are assigned to multiple hubs in order to further reduce cost as compared to the single allocation case.

4.6 SOLUTION TECHNIQUES FOR VRPTW

A solution of the VRPTW consists of a set of routes covering the demand of all customers (as shown in Figure 4.1). Each vehicle route in turn consists of a chain of customers visited in a pre-specified order for time feasibility and other constraints mentioned in Section 4.2.1. As many feasible such chains can be constructed, the VRPTW is considered as a combinatorial optimisation problem. These chains of customers and the routes they constitute can be represented in variety of ways in the solution algorithm. For example, Figure 4.2 shows a solution representation for a genetic algorithm (such as

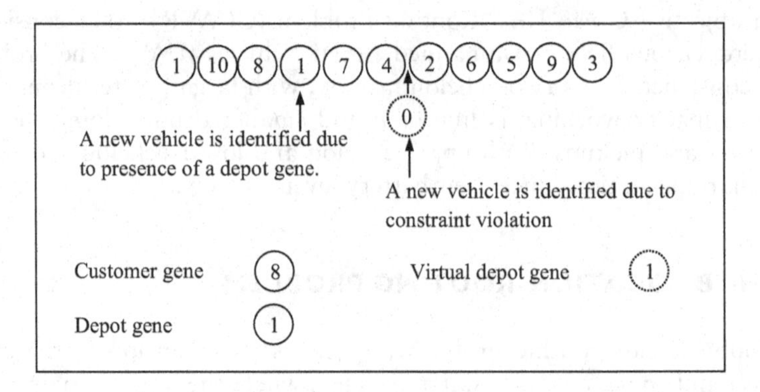

Figure 4.2 A representation of a VRPTW solution in genetic algorithm.

the one used in Qureshi et al., 2009b), where the complete solution is represented as a single string of customers and depot(s) and which always returns a feasible solution. Many different ways can be used to represent solutions of VRPs with additional side conditions, for example, Qureshi and Taniguchi (2017) used a double layer representation where one layer represented the customers' chain, whereas a synchronised layer contained the amount of delivery goods for the customers of layer one. Each layer represented a separate decision variable, *i.e.*, the visiting sequence and delivery amount.

4.6.1 Heuristics solution approaches

Since its first appearance in 1959 (Dantzig & Ramser, 1959), many heuristics or approximate solution approaches have been proposed for the VRP and its extensions. Heuristics techniques are procedures that seek to find good (*i.e.*, near exact) solutions for mathematical programming problems at a reasonable computational cost. They do not guarantee to identify the exact solution or state how close to the exact solution a particular feasible solution is (Thompson & van Duin, 2003). Heuristics, which are classified as route building algorithms such as the insertion heuristics (Solomon, 1987) start from empty routes and add customers to the route based on some criteria (such as minimum insertion cost) until no further addition is possible. Route improvement algorithms on the other hand try to improve currently available routes. The saving algorithm presented by Clarke and Wright (1964) could be classified as one of the simplest route improving algorithms, which starts with a separate route for each customer and then merges these based on the resulting savings in distance travelled. More advanced route improvement techniques (*e.g.*, see Savelsbergh, 1992) exchange the position of a single customer (or a sub-chain of customers) within the same route or between two routes (this operation is called exploration of the neighbourhood of the current solution (routes) $N(S)$).

Metaheuristics embed neighbourhood search in an iterative framework, for example, tabu search (TS) (Garcia et al. 1994) iteratively searches the neighbourhood $N(S)$ of a current solution S by accepting the better solution. To avoid cycling, a tabu list is maintained that stores recently visited paths and other forbidden moves for a duration of tabu tenure (number of iterations). The presence of a tabu list also allows heuristics to accept even a poorer solution as the current solution (thus providing a possibility of avoiding convergence to a local minimum solution). The tabu list can be overridden if some of the aspiration criteria are met such as obtaining a new global best solution.

Genetic algorithms (GA) are another popular metaheuristic type that is often used for solving the VRP and its variants (*e.g.*, see Tan et al., 2001). Instead of just improving a single solution, GA creates a population of many solutions, where each solution is called a chromosome and each customer node is termed a gene (as shown in Figure 4.2). GA explores the neighbourhoods of these solutions by exchanging sub-sequences of customers' genes between parent chromosomes (termed as crossover) and creating new child chromosomes, which then compete (based on their fitness such as better solution cost) with the parents to survive in the next generation of population for the next iteration. Mutation (such as random swap of customers) is used to provide the possibility of expanding the search space to avoid a local minimum.

Many other heuristics have also been proposed based on evolutionary or mimicking natural phenomenon such as Ants routing, particle swarm algorithms, neural networks and so on. Most of these work on the ideas of exploration of the neighbourhood of the current solution(s) (such as crossover) and on the use of some strategy to widen the search space and avoid local minimums (such as mutation).

4.6.2 Exact solution approaches

Exact solution techniques solve the mathematical model of the VRPTW (4.1–4.9) in a structured way to obtain the global minimum solution. For example, it can be observed that only constraint (4.2) links all the vehicles (routes) of a feasible VRPTW solution, whereas rest of the constraints can be separated for each vehicle route making it a single vehicle problem (such as a shortest path problem). This structure is exploited in the exact solution. For example, the *Dantzig-Wolfe decomposition* (Dantzig & Wolfe, 1960) of the VRPTW results in the set partitioning master problem and an Elementary Shortest Path Problem with Resource Constraints (ESPPRC) as its subproblem, for every vehicle $k \in K$ (Qureshi et al., 2009a). The master problem, which consists of selecting a set of feasible paths of minimum cost (keeping the linking constraint of the original formulation), is mathematically described as:

$$\min \sum_{p \in P} c_p y_p \tag{4.10}$$

subject to

$$\sum_{p \in P} a_{ip} y_p = 1 \qquad \forall i \in C \tag{4.11}$$

$$y_p \in \{0, 1\} \qquad \forall p \in P \tag{4.12}$$

Here, the set of all feasible paths is given by P and y_p takes value 1 if the path $p \in P$ is selected and 0 otherwise. The cost of the path p is denoted by c_p, and a_{ip} represents the number of times path p serves customer i. Theoretically set P could be very large depending on the number of customers n; therefore, the Reduced Master Problem (RMP) is solved based on the available partial set of feasible paths ($\tilde{P}$). Also, in actual application the set covering master problem is solved by replacing the equal sign in (Eq. 4.11) with equal or greater sign constraint, as the linear programming relaxation of set covering type master problem is more stable than the set partitioning type (Desrochers et al., 1992). As far as the cost matrix c_{ij} follows the triangular inequality, even the set covering formulation of the VRPTW contains the *integrality property*, and hence an optimum solution would only visit each customer once.

Column generation starts with a feasible *basis* and the RMP is solved which provides the current minimum value $Z_{LP}(\tilde{P})$, where $\tilde{P}$ represents the initial set of columns available in the master problem's Linear Programme (LP). The objective function of the ESPPRC subproblem (*pricing problem* or *pricing oracle*) is given by Eq. (4.13), subject to the constraints represented by Eqs. (4.3–4.9). The reduced cost matrix for the subproblem is obtained using Eq. (4.14); where π_i are the dual variable values (*prices* or *simplex multipliers*) corresponding to the assignment constraint (Eq. 4.11) in the master problem. The role of the subproblem is to produce feasible paths with a negative reduced cost (*marginal cost*). During every column generation iteration, the cyclic solutions of RMP and the subproblem are repeated and the partial set $\tilde{P}$ is kept on augmented with the paths (*columns*) generated in the subproblem. The column generation schemes stop, giving the optimum solution when the subproblem fails to return a column with a negative reduced marginal cost.

$$\min \sum_{(i,j) \in A} \overline{c_{ij}} x_{ijk} \tag{4.13}$$

$$\overline{c_{ij}} = c_{ij} - \pi_i \qquad \forall\, i \in V \tag{4.14}$$

Another popular set of exact algorithms utilises the concept of Lagrangian relaxation of hard constraints (*e.g.*, see Kohl & Madsen, 1997). The

Lagrangian relaxation of the assignment constraint (4.2) results in the relaxed (dual) objective function given by Eq. (4.15).

$$\min \sum_{k \in K} \sum_{i \in V} \sum_{j \in V} (c_{ij} - \lambda_j) x_{ijk} + \sum \lambda_i \tag{4.15}$$

subject to constraints given by Eqs. (4.3–4.9).

For any value of the Lagrangian multiplier vector $\lambda = (\lambda_1, \lambda_2, \ldots, \lambda_{|C|})$, the resulting problem is an Elementary Shortest Path Problem with Time Windows and Capacity Constraint (ESSPTWCC), for each vehicle.

4.7 CONCLUSIONS

The history of research on optimisation of various components of urban logistics is very old, and the trend of introducing new problems by making them closer to reality and of finding new innovative algorithms for finding their solutions is still growing. New technologies such as electric vehicles, drones and delivery robots have inspired researchers towards a lot of new variants in the vehicle routing problem. The number and variety of green vehicle routing models are increasing due to the increased sensitivity towards the environmental footprint of urban freight and due to increased commitments from various governments towards carbon neutral cities. Models which try to optimise multiple aspects of the city logistics simultaneously (such as location of logistics facilities at all levels and routing among them) are also gaining importance and more sophisticated algorithms are being designed to solve them in reasonable amount of time so that their acceptance in the practice can be increased.

REFERENCES

Afroditi, A., Boile, M., Theofanis, S., Sdoukopoulos, E. and Margaritis, D. (2014). Electric vehicle routing problem with industry constraints: Trends and insights for future research, *Transportation Research Procedia*, 3, 452–459.

Afshar-Bakeshloo, M., Mehrabi, A., Safari, H., Maleki, M. and Jolai, F. (2016). A green vehicle routing problem with customer satisfaction criteria, *Journal of Industrial Engineering International*, 12, 529–544.

Ahani, P., Arantes, A. and Melo, S. (2016). A portfolio approach for optimal fleet replacement toward sustainable urban freight transportation, *Transportation Research Part D*, 48, 357–368.

Akca, Z., Berger, R. and Ralphs, T. (2009). A branch-and-price algorithm for combined location and routing problems under capacity restrictions, In: Chinneck, J.W., Kristjansson, B. and Saltzman, M.J. (eds.), *Operations Research and Cyber-Infrastructure*, Springer, USA, 309–330.

Anderluh, A., Nolz, P.C., Hemmelmayr, V.C. and Crainic, T.G. (2021). Multi-objective optimization of a two-echelon vehicle routing problem with vehicle synchronization and "grey zone" customers arising in urban logistics, *European Journal of Operational Research*, **289**, 940–958.

Basso, R., Kulcsár, B., Egardt, B., Lindroth, P. and Sanchez-Diaz, I. (2019). Energy consumption estimation integrated into the electric vehicle routing problem, *Transportation Research Part D*, **69**, 141–167.

Bektas, T. and Laporte, G. (2011). The pollution-routing problem, *Transportation Research Part B*, **45**, 1232–1250.

Belengure, J., Benavent, E. and Prins, C. (2011). A branch-and-cut method for the capacitated location-routing problem, *Computers & Operations Research*, **38**, 931–941.

Bertsimas, D.J., Jaillet, P. and Odoni, A.R. (1990). A priori optimization, *Operations Research*, **38**, 1019–1033.

Bhusiri, N., Qureshi, A.G. and Taniguchi, E. (2014). The trade off between fixed vehicle costs and time-dependent arrival penalties in a routing problem, *Transportation Research Part E*, **62**, 1–22.

Bostel, N., Dejax, P. and Zhang, M. (2015). A model and a metaheuristic method for the hub location routing problem and application to postal services. In 2015 *International Conference on Industrial Engineering and Systems Management* (IESM). IEEE, 1383–1389.

Branchini, R.M., Armentano, V. A. and Løkketangen, A. (2009). Adaptive granular local search heuristics for a dynamic vehicle routing problem, *Computers and Operations Research*, **36**, 2955–2968.

Breunig, U., Baldacci, R., Hartl, R.F. and Vidal, T. (2019). The electric two-echelon vehicle routing problem, *Computers and Operations Research*, **103**, 198–210.

Campbell, J. F. and O'Kelly, M.E. (2012). Twenty-five years of hub location research, *Transportation Science*, **46**, 153–169.

Çetiner, S., Sepil, C. and Süral, H. (2010). Hubbing and routing in postal delivery systems, *Annals of Operations research*, **181**, 109–124.

Chen, Z. and Xu, H. (2006). Dynamic column generation for dynamic vehicle routing with time windows, *Transportation Science*, **40**, 74–88.

Choi, E., and Tcha, D.W. (2007). A column generation approach to the heterogeneous fleet vehicle routing problem, *Computers & Operations Research*, **34**, 2080–2095.

Christiansen, C.H. and Lysgaard, J. (2007). A branch-and-price algorithm for the capacitated vehicle routing problem with stochastic demands, *Operations Research Letters*, **35**, 773–781.

Clarke, G. and Wright, J.W. (1964). Scheduling of vehicles from a central depot to a number of delivery points, *Operations Research*, **1**, 568–581.

Coelho, B.N., Coelho, V. N., Coelho, I.M., Ochi, L.S., Haghnazar, R.K., Zuidema, D., Lima, M.S.F. and da Costa, A.R. (2017). A multi-objective green UAV routing problem, *Computers and Operations Research*, **88**, 306–315.

Conrad, R.G. and Figliozzi, M.A. (2011). The recharging vehicle routing problem, In: Doolen, T. and Van Aken, E. (eds.), *Proceedings Industrial Engineering Research Conference*, Reno, NV. https://www.researchgate.net/publication/267202427_The_Recharging_Vehicle_Routing_Problem

Crainic, T.G., S. Mancini, G. Perboli and R. Tadei. (2008). Clustering-based heuristics for the two-echelon vehicle routing problem, Interuniversity Research Centre on Enterprise Networks, Logistics and Transportation (CIRRELT). https://www.cirrelt.ca/documentstravail/cirrelt-2008-46.pdf

Dantzig, G.B. and Wolfe, P. (1960). Decomposition principle for linear programs, *Operations Research*, **8**, 101–111.

Dantzig, G.B. and Ramser, J.H. (1959). The truck dispatching problem, *Management Science*, **6**, 80–91.

Davis, B. A. and Figliozzi, M. A. (2013). A methodology to evaluate the competitiveness of electric delivery trucks, *Transportation Research Part E*, **49**, 8–23.

Deng, P., Amirjamshidi, G. and Roorda, M. (2020). A vehicle routing problem with movement synchronization of drones, sidewalk robots, or foot-walkers, *Transportation Research Procedia*, **46**, 29–36.

Desrochers, M., Desrosiers, J. and Solomon, M. (1992). A new optimization algorithm for the vehicle routing problem with time windows, *Operations Research*, **40**, 342–354.

Dror, M. and Trudeau, P. (1986). Stochastic vehicle routing with modified savings algorithm, *European Journal of Operational Research*, **23**, 228–235.

Dukkanci, O., Kara, B.Y. and Bektas, T. (2019). The green location-routing problem, *Computers and Operations Research*, **105**, 187–202.

Fiori, C. and Marzano, V. (2018). Modelling energy consumption of electric freight vehicles in urban pickup/delivery operations: analysis and estimation on a real-world dataset, *Transportation Research Part D*, **65**, 658–673.

Fleischmann, B., Gnutzmann, S. and Sandvoss, E. (2004). Dynamic vehicle routing based on online traffic information, *Transportation Science*, **38**, 420–433.

Garcia, B.L., Potvin, J.Y. and Rousseau, J.M. (1994). A parallel implementation of the Tabu search heuristic for vehicle routing problem with time windows constraints, *Computers and Operations Research*, **21**, 1025–1033.

Gendreau, M., Laporte, G. and Seguin, R. (1996a). A Tabu search heuristics for the vehicle routing problem with stochastic demands and customers, *Operations Research*, **44**, 469–477.

Gendreau, M., Laporte, G. and Seguin, R. (1996b). Stochastic vehicle routing, *European Journal of Operational Research*, **88**, 3–12.

Gendreau, M., Laporte, G. and Seguin, R. (1995). An exact algorithm for the vehicle routing problem with stochastic demands and customers, *Transportation Science*, **29**, 143–155.

Ghiani, G., Laporte, G. and Musmanno, R. (2013). *Introduction to Logistics Systems Management*, John Wiley & Sons, Ltd, West Sussex.

Giordano, A., Fischbeck, P. and Matthews, H.S. (2018). Environmental and economic comparison of diesel and battery electric delivery vans to inform City Logistics fleet replacement strategies, *Transportation Research Part D*, **64**, 216–229.

Goeke, D. and Schneider, M. (2015). Routing a mixed fleet of electric and conventional vehicles, *European Journal of Operational Research*, **245**, 81–99.

Gonzalez-Feliu, J. (2013). Vehicle routing in multi-echelon distribution systems with cross-docking: A systematic lexical-metanarrative analysis, *Computer and Information Science*, **6**, 28–47.

Gonzalez-Feliu, J. (2008). *Models and Methods for the City Logistics: The Two-Echelon Vehicle Routing Problem.* PhD Thesis, Politecnico di Torino, Turin, Italy.

Hof, J., Schneider, M. and Goeke, D. (2017). Solving the battery swap station location-routing problem with capacitated electric vehicles using an AVNS algorithm for vehicle-routing problems with intermediate stops, *Transportation Research Part B*, **97**, 102–112.

Jacobsen, S.K. and Madsen, O.B. (1978). On the location of transfer points in a two-level newspaper delivery system - A case study, *The International Symposium on Location Decisions*, Alberta, Canada, 24–28.

Jeeninga, H., van Arkel, W.G. and Volkers, C.H. (2002). *Performance and Acceptance of Electric and Hybrid Vehicles*, Municipality of Rotterdam, Rotterdam, the Netherlands.

Jia, S., Deng, L., Zhao, Q. and Chen, Y. (2022). An adaptive large neighborhood search heuristic for multi-commodity two-echelon vehicle routing problem with satellite synchronization, *Journal of Industrial & Management Optimization*, **19**, 1187–1210.

Jie, W., Yang, J., Zhang, M. and Huang, Y. (2019). The two-echelon capacitated electric vehicle routing problem with battery swapping stations: Formulation and efficient methodology, *European Journal of Operational Research*, **272**, 879–904.

Jabali, O., Van Woensel, T. and De Kok, A. (2012). Analysis of travel times and CO_2 emissions in time-dependent vehicle routing, *Production and Operations Management*, **21**, 1060–1074.

Karimi, H. (2018). The capacitated hub covering location-routing problem for simultaneous pickup and delivery systems, *Computers & Industrial Engineering*, **116**, 47–58.

Keskin, M., Laporte, G. and Catay, B. (2019). Electric vehicle routing problem with time-dependent waiting times at recharging stations, *Computers and Operations Research*, **107**, 77–94.

Koc, C. (2019). Analysis of vehicle emissions in location-routing problem, *Flexible Services and Manufacturing Journal*, **31**, 1–33.

Kohl, N. and Madsen, O.B.G. (1997). An optimization algorithm for the vehicle routing problem with time windows based on Lagrangian relaxation, *Operations Research*, **45**, 395–406.

Kramer. R., Subramanian, A., Vidal, T. and Cabral, L.A.F. (2015). A matheuristic approach for the pollution-routing problem, *European Journal of Operational Research*, **243**, 523–539.

Kuo, Y. and Wang, C. (2011). Optimizing the VRP by minimizing fuel consumption, *Management of Environmental Quality: An International Journal*, **22**, 440–450.

Laporte, G., Louveaux, F. and Mercure, H. (1992). The vehicle routing problem with stochastic travel times. *Transportation Science*, **26**, 161–170.

Laporte, G. and Dejax, P.J. (1989). Dynamic location-routeing problems, *Journal of the Operational Research Society*, **40**, 471–482.

Laporte, G., Louveaux, F. and Mercure, H. (1989). Models and exact solutions for a class of stochastic location-routing problems, *European Journal of Operational Research*, **39**, 71–78.

Laporte, G. and Nobert, Y. (1981). An exact algorithm for minimizing routing and operating costs in depot location, *European Journal of Operational Research*, 6, 224–226.

Larsen, A., Madsen, O. and Solomon, M. (2002). Partially dynamic vehicle routing – Models and algorithms. *Journal of Operation Research Society*, 53, 637–646.

Lebeau, P., Macharis, C. and van Mierlo, J. (2016). Exploring the choice of battery electric vehicles in city logistics: A conjoint-based choice analysis, *Transportation Research Part E*, **91**, 245–258.

Lebeau, P., De Cauwer, C., VanMierlo, J., Macharis, C., Verbeke, W. and Coosemans, T. (2015). Conventional, hybrid, or electric vehicles: Which technology for an urban distribution centre? *The Scientific World Journal*, **2015**, **1–11**.

Lee, D.Y., Thomas, V.M. and Brown, M.A. (2013). Electric urban delivery trucks: Energy use, greenhouse gas emissions, and cost-effectiveness, *Environmental Science & Technology*, **47**, 8022–8030.

Li, H., Zhang, L., Lv, T. and Chang, X. (2016). The two-echelon time-constrained vehicle routing problem in linehaul-delivery systems, *Transportation Research Part B: Methodological*, **94**, 169–188.

Lin, J. and Lei, H. (2009). Distribution systems design with two-level routing considerations, *Annals of Operations Research*, **172**, 329–347.

Lin, C., Choy, K.L., Ho, G.T.S., Chung, S.H. and Lam, H.Y. (2014). Survey of green vehicle routing problem: Past and future trends, *Expert Systems with Applications*, **41**, 1118–1138.

Macharis, C., Lebeau, P., van Mierlo, J. and Lebeau, K. (2013). Electric versus conventional vehicles for logistics: A total cost of ownership, *EVS27*, November 17–20, 2013, Barcelona, Spain.

Montoya, A., Gueret, C., Mendoza, J.E. and Villegas, J.G. (2017). The electric vehicle routing problem with nonlinear charging function, *Transportation Research Part B*, **103**, 87–110.

Nikbakhsh, E. and Zegordi, S. (2010). A heuristic algorithm and a lower bound for the two-echelon location-routing problem with soft time window constraints, *Scientia Iranica Transaction E: Industrial Engineering*, **17**, 36–47.

Ponboon, S., Qureshi, A.G. and Taniguchi, E. (2016). Branch-and-price algorithm for the location-routing problem with time windows, *Transportation Research Part E*, **86**, 1–19.

Psaraftis, H. N. (1995). Dynamic vehicle routing: Status and prospects, *Annals of Operational Research*, **61**, 143–164.

Qureshi, A. G. and Taniguchi, E. (2017). A multi-period relief distribution model considering limited resources and decreasing resilience of affected population, *Journal of the Eastern Asia Society for Transportation Studies*, **12**, 57–73.

Qureshi, A.G., Taniguchi, E. and Yamada, T. (2012). A microsimulation based analysis of exact solution of dynamic vehicle routing with soft time windows, *Procedia - Social and Behavioral Sciences*, **39**, 205–216.

Qureshi, A.G., Taniguchi, E. and Yamada, T. (2009a). An exact solution approach for vehicle routing and scheduling problems with soft time windows, *Transportation Research Part E*, **45**, 960–977.

Qureshi, A.G., Taniguchi, E. and Yamada, T. (2009b). Hybrid insertion heuristics for vehicle routing and scheduling problems with soft time windows, *Infrastructure Planning Review*, **26**, 703–714.

ReVelle, C., Cohon, J. and Shobrys, D. (1991). Simultaneous siting and routing in the disposal of hazardous wastes, *Transportation Science*, **25**(2), 138–145.

Salhi, S. and Nagy, G. (1999). Consistency and robustness in location-routing, *Studies in Locational Analysis*, **13**, 3–19.

Savelsbergh, M.W.P. (1992). The vehicle routing problem with time windows: Minimizing route duration, *ORSA Journal on Computing*, **4**, 146–154.

Schiffer, M., Schneider, M. and Laporte, G. (2018). Designing sustainable mid-haul logistics networks with intra-route multi-resource facilities, *European Journal of Operational Research*, **265**, 517–532.

Sen, B., Ercan, T. and Tatari, O. (2017). Does a battery-electric truck make a difference? - Life cycle emissions, costs, and externality analysis of alternative fuel-powered Class 8 heavy-duty trucks in the United States, *Journal of Cleaner Production*, **141**, 110–121.

Solomon, M.M. (1987). Algorithms for the vehicle routing and scheduling problem with time windows constraints, *Operations Research*, **35**, 254–265.

Srivastava, R. and Benton, W.C. (1990). The location-routing problem: Consideration in physical distribution system design, *Computers and Operations Research*, **6**, 427–435.

Stewart, W. R. Jr. and Golden, B.L. (1983). Stochastic vehicle routing: A comprehensive approach, *European Journal Operational Research*, **14**, 371–385.

Tan, K.C., Cheong, C.Y. and Goh, C.K. (2007). Solving multiobjective vehicle routing problem with stochastic demand via evolutionary computation, *European Journal Operational Research*, **177**, 813–839.

Tan, K.C., Lee, L.H. and Ou, K. (2001). Artificial intelligence heuristics in solving vehicle routing problems with time window constraints, *Engineering Applications of Artificial Intelligence*, **14**, 825–837.

Taniguchi, E. and Shimamoto, H. (2004). Intelligent transportation system based dynamic vehicle routing and scheduling with variable travel time, *Transportation Research - Part C*, **12**, 235–250.

Taniguchi, E., Thompson, R.G., Yamada, T. and van Duin, R. (2001). *City logistics; Network Modeling and Intelligent Transport Systems*, Pergamon, Oxford.

Thompson, R.G. and van Duin, R. (2003). Vehicle routing and scheduling, In: Taniguchi, E. and Thompson, R.G. (eds.), *Innovations in Freight Transportation*, WIT Press, Southampton, 47–63.

van Duin, J.H.R., Tavasszy, L.A. and Quak, H.J. (2013). Towards E(lectric)-urban freight: First promising steps in the electric vehicle revolution, *European Transport*, **54**, 1–19.

Vermie, T. (2002), *Electric Vehicle City Distribution*, Final Report, European Commission Project.

Wang, Y., Assogba, K., Liu, Y., Ma, X.L., Xu, M.Z. and Wang, Y.H. (2018). Two-echelon location-routing optimization with time windows based on customer clustering, *Expert Systems with Applications*, **104**, 244–260.

Wu. Y., Qureshi, A.G. and Yamada, T. (2022). Adaptive large neighborhood decomposition search algorithm for multi-allocation hub location routing problem, *European Journal of Operations Research*, **302**, 1113–1127.

Xiao, Y., Zhao, Q., Kaku, I. and Xu, Y. (2012). Development of a fuel consumption optimization model for the capacitated vehicle routing problem, *Computers and Operations Research*, **39**, 1419–1431.

Zografos, K. G. and Samara, S. (1989). Combined location-routing model for hazardous waste transportation and disposal, *Transportation Research Record*, **1245**, 52–59.

Chapter 5

Multi-agent simulation with machine learning

5.1 INTRODUCTION

Multi-agent modelling provides a useful tool for representing and understanding the behaviour of stakeholders, *i.e.*, shippers, freight carriers, administrators, and residents who are involved in city logistics. Collaboration between stakeholders is essential for successful operation of city logistics and multi-agent simulation allows prediction of the effects of city logistics policy measures before implementing them in real urban areas and to find appropriate management for satisfying the objectives of stakeholders. Smart city logistics are based on advanced information systems and this aspect can be incorporated in multi-agent simulation.

Machine learning which includes Q-learning and other reinforcement learning methods can be used in multi-agent simulation for replicating the choice behaviour of stakeholders of options in city logistics. For example, freight carriers can choose to deliver through urban consolidation centres (UCC) or directly to the customers to minimise their distribution costs and this type of behaviour can be modelled using machine learning. Multi-agent simulation with machine learning presents a powerful method to model the behaviour of stakeholders for finding better solutions in city logistics.

In addition to modelling the behaviour of stakeholders, multi-agent simulation can help multi-stakeholder decision making in the framework of public-private partnerships (PPP) of city logistics. The multi-agent simulation is able to consider responses of private companies to policy measures by public authorities and also interactions between stakeholders. Therefore, evaluating policy measures with multiple criteria is possible using the multi-agent simulation before implementing them.

5.2 MULTI-AGENT MODELLING

Agent models are composed of intelligent agents and the environment. The agents perceive the state of the environment and act on the state of the environment based on the calculation executed using intelligent algorithms

DOI: 10.1201/9781003261513-6

within the agents. Multi-agent models have multiple agents, for example, shippers, freight carriers, administrators, residents, and UCC operators in city logistics. These autonomous agents can make decisions and interact with each other based on learning data to achieve their objectives. The multi-agent simulation can duplicate the process of interaction among agents.

Firdausiyah et al. (2019) presented multi-agent simulation models using adaptive dynamic programming which is a reinforcement learning method. In multi-agent models each agent learns to maximise the rewards which are given based on the decision of an agent. They applied these models in cases of urban goods delivery using urban consolidation centres (UCC) in a hypothetical road network. They considered the interaction among stakeholders including freight carriers, UCC operator, customers, administrators, and residents who are involved in city logistics. They assumed freight carriers and the UCC operator are learning agents. The UCC operator offers a fee for using Joint Delivery Systems (JDS) through the UCC and freight carriers make decisions if they use JDS or Direct Delivery (DD) to customers based on their costs. The UCC operator uses the reinforcement learning system for maximising their profits and freight carriers use the UCC for minimising the delivery costs.

The performance of two methods of reinforcement learning: Adaptive Dynamic Programming (ADP)–based reinforcement learning and Q-learning were evaluated (Firdausiyah et al., 2019). The ADP-based reinforcement learning employs an on-policy algorithm which attempts to evaluate and improve the policy that is used to make decisions, while Q-learning is an off-policy algorithm which learns the value of the optimal policy independently of the agent's actions (Sutton & Barto, 1998). Q-learning is one of the Temporal Difference (TD) algorithms that directly learns the action value function (Q) instead of the state value function (V) and thus does not require a model of the environment (Watkins & Dayyan, 1992). The learned action value function in Q-learning directly approximates the optimal action value function (Q*) as the maximum possible state-action value, independently being followed by the policy. Thus Q-learning is considered to be an off-policy algorithm.

In contrast to Q-learning, ADP-based reinforcement learning has a separate memory structure to represent a policy and value function. It consists of three networks: A critic network, a policy network, a model network. The ADP-based reinforcement learning architecture incorporates the relationship of policy as an actor, value function as a critic, and the model given by the environment as a prediction model to estimate events in the next state in the ADP-based reinforcement model. The ADP-based reinforcement model as an on-policy learning model evaluates and improves the policy that is used for decision making.

Firdausiyah et al. (2019) showed the change in profit of a UCC operator using the ADP-based reinforcement learning and Q-learning. The results indicate that the ADP-based reinforcement learning outperformed Q-learning by 10.7% at the end of simulations for 168 days in cases of fluctuating demand of customers. The evaluation using multiple criteria was also done considering reducing delivery costs and CO_2, NOx, and SPM emissions.

Rabe et al. (2018) applied a discrete event simulation to evaluate a UCC for distribution in Athens. They concluded that the simulation showed that utilising UCCs lowered the number of tours needed to meet the customer's demands and decreased the required distance travelled by freight vehicles.

Joubert (2018) applied the multi-agent transport simulation (MATSim) for evaluating the relocation of an urban container terminal in large scale scenarios in Cape Town, South Africa. Results showed that the simulation provided insights about how specific operators and the public at large will be influenced, both from an overall vehicle-kilometres travelled point of view and the change in travel times.

Jlassi et al. (2017) evaluated regulatory policies including the vehicle size and time windows in city logistics using a multi-agent simulation model in Paris. They tested four scenarios and discussed the results from various viewpoints, including the number of vehicles, emissions of CO_2, NOx, and SO_2.

Taniguchi et al. (2018) demonstrated how multi-agent simulation can be used for evaluating a combination of city logistics policy measures focusing on consolidation centres, green management and parking management. They pointed out that in public-private partnerships (PPP), multi-agent models will play an important role for evaluating city logistics policy measures based on data. Sharing data from both the public and private sectors is essential and interaction between both sectors is necessary to find better solutions using the results provided by multi-agent simulation.

Sakai et al. (2020) presented an agent-based urban freight simulation platform to evaluate policy measures in city logistics. The simulator is capable of simulating commodity contracts, logistics and vehicle operation planning, and parking decisions in a fully disaggregated manner. They applied their models in Singapore and revealed network level impacts of night-time and off-peak deliveries of goods.

5.3 DECISION SUPPORT SYSTEMS

Multi-agent models can be used for multi-stakeholder decision making based on this modelling approach. Le Pira et al. (2017) applied integrated models combining discrete choice models (DCM) with agent-based models (ABM). These models consider stakeholders' heterogeneous preferences and simulate their interactive behaviour in a consensus building process. They pointed out that the integrated models of DCM and ABM can overcome their weaknesses, since it is grounded on sound microeconomic theory providing detailed (static) stakeholders' behavioural knowledge, but it is also capable of reproducing agents' (dynamic) interaction during the decision-making process. Lebeau et al. (2018) presented the multi-actor multi-criteria analysis (MAMCA) (see Chapter 7) as a method which structured the consultation process of stakeholders in city logistics. They tested this approach in a workshop in Brussels and concluded that the methodology allowed authorities to identify the priorities of the stakeholders in city logistics, to guide the discussion towards a consensus and to provide inputs for an eventual adoption of strategies. Gatta et al. (2019) applied an interactive MAMCA for analysing off-hour delivery (OHD) solutions with stakeholders in Rome. They pointed out that stakeholders prefer a solution where OHD are jointly implemented together with one or more urban consolidation centres.

5.4 CONCLUSIONS

Multi-agent modelling with machine learning such as reinforcement learning provides a useful tool for duplicating the behaviour of stakeholders in city logistics. Multi-agent simulation is able to help multi-stakeholder decision making in the framework of public-private partnerships for choosing appropriate policy measures before implementing them. Multi-actor multi-criteria analysis (MAMCA) can structure a consultation process for stakeholders.

REFERENCES

Firdausiyah, N., Taniguchi, E. and Qureshi, A.G. (2019). Modeling city logistics using adaptive dynamic programming based multi-agent simulation, *Transportation Research Part E*, **125**, 74–96.

Gatta, V., Marcucci, E., Site, P. D., Le Pira, M. and Carrocci, C. S. (2019). Planning with stakeholders: Analysing alternative off-hour delivery solutions via an interactive multi-criteria approach, *Research in Transportation Economics*, **73**, 53–62.

Jlassi, S., Tamayo, S., Gaudron, A. and de La Fortelle, A. (2017). Simulating impacts of regulatory policies on urban freight: Application to the catering setting, *6th IEEE International Conference on Advanced Logistics and Transport (ICALT)*, Bali, Indonesia, 106–112.

Joubert, J. (2018). Evaluating the relocation of an urban container terminal, In: E. Taniguchi and R. G. Thompson (eds.) *City Logistics 2*, ISTE-Wiley, London, 197–210.

Le Pira, M., Marcucci, E., Gatta, V., Ignaccolo1, M., Inturri, G. and Pluchino, A. (2017). Towards a decision-support procedure to foster stakeholder involvement and acceptability of urban freight transport policies, *European Transportation Research Review*, **9**, 54.

Lebeau, P., Cathy Macharis, C., Van Mierlo, J. and Janjevic, M. (2018). Improving policy support in city logistics: The contributions of a multi-actor multi-criteria analysis, *Case Studies on Transport Policy*, **6**, 554–563.

Rabe, M., Klueter, A. and Wuttke, A. (2018). Evaluating the consolidation of distribution flows using a discrete event supply chain simulation tool: Application to a case study in Greece, *Proceedings of the IEEE 2018 Winter Simulation Conference*, Gothenburg.

Sakai, T., Alho, A.R., Bhavathrathan, B.K., Chiara, G.D., R.G., Jinge, P., Hyodo, T., Cheah, L., Ben-Akiva, M. (2020). SimMobility freight: An agent-based urban freight simulator for evaluating logistics solutions, *Transportation Research Part E*, **141**, 102017.

Sutton, R.S., Barto, A.G., (1998). *Reinforcement Learning: An Introduction,* The MIT Press, Cambridge, MA, London.

Taniguchi, E., Qureshi, A. G. and Konda, K. (2018). Multi-agent simulation with reinforcement learning for evaluating a combination of city logistics policy measures, In: Taniguchi, E. and Thompson, R.G. (eds), *City Logistics 2: Modelling and Planning Initiatives*, ISTE, London, 165–178.

Watkins, C.J.C.H., and Dayyan, P. (1992). Q-learning. *Machine Learning*, **8**, 279–292.

Chapter 6

Reliability and resilience

6.1 INTRODUCTION

There have been increasing concerns about risks caused by natural hazards including earthquakes, flooding, tsunamis, snowfalls and bush fires as well as manmade hazards of crashes and terrorist attacks. Although in principle these risks should be well assessed and incorporated in city logistics, they are not fully taken into account in modelling city logistics (Taniguchi et al., 2001) and implementing city logistics schemes in urban areas. The reasons are: (a) assessing the risks related to city logistics is hard due to the uncertainty of these events, (b) incorporating risks of natural and manmade hazards incurs additional costs on logistical operation, and (c) natural and manmade disasters are not regarded as being within the logistics managers' responsibility.

Recently we have encountered extremely destructive disasters generated by the tsunami after Northern Sumatra earthquake in Indian Sea region in 2004, Hurricane Catharina in Mexican Bay area, USA in 2005, Sichuan earthquake in China in 2008 and bush fires in Melbourne, Australia in 2009, Tohoku earthquake in Japan in 2011. As well, we had a chemical attack using saran on the subways in Tokyo in 1995, September 11 terrorism in New York in 2001, a blast attack in London in 2005 and a piracy attack off the Somalia coast in 2009. These threats triggered a change of mindsets of stakeholders to start taking into account risks due to natural and manmade hazards.

In private firms these risks are discussed from the viewpoint of business continuity planning. We can find a typical case of critical disruption of supply chain caused by the collapse of a small firm named Riken by the Niigataken Chuetsu-oki earthquake in 2007, which produced piston rings for automobile manufacturers in Japan. The disruption of the factory of Riken generated a shutdown of many automobile assembly factories for a week since small inventory levels of such parts in Just-In-Time production systems were vulnerable against the threats of the earthquake and there was too much dependence on supplying a critical part of a single firm. After that many logistics companies realised the importance of incorporating

DOI: 10.1201/9781003261513-7

risks of natural disasters in logistics systems and considered business continuity planning.

In the public sector risks due to natural and manmade hazards in urban freight transport systems are directly related to public welfare and public health in emergency situations. The public sector is interested in how to mitigate the damage to urban logistics facilities and recover quickly after disasters from the viewpoint of delivering goods needed for maintaining a high quality of life in urban areas.

In this context, we need to consider multiple stakeholders involved in urban freight transport systems, namely shippers, freight carriers, administrators and residents (consumers) to incorporate risks in city logistics. As there are different motivations, objectives and behaviour among stakeholders when facing risks, special methodologies including multi-agent models are required for modelling the behaviour of stakeholders.

In general, city logistics takes into account the day-to-day risks of delays when arriving at customers for delivering or picking up goods due to recurrent congestion generated by the concentration of traffic during peak periods, crashes and sports events. However, we need to incorporate less frequent and severer effects caused by extreme events such as cyclones, earthquakes and flooding. Figure 6.1 presents a classification of risks related to city logistics. The horizontal axis shows the source of difficulty for assessing risks caused by events and the vertical axis indicates the frequency of events.

The first level of difficulty arises from complexity. The effects of congestion on road networks are complex and it is hard to anticipate the travel times due to congestion since the demand of passenger and freight traffic are fluctuating. Conventional vehicle routing and scheduling with time

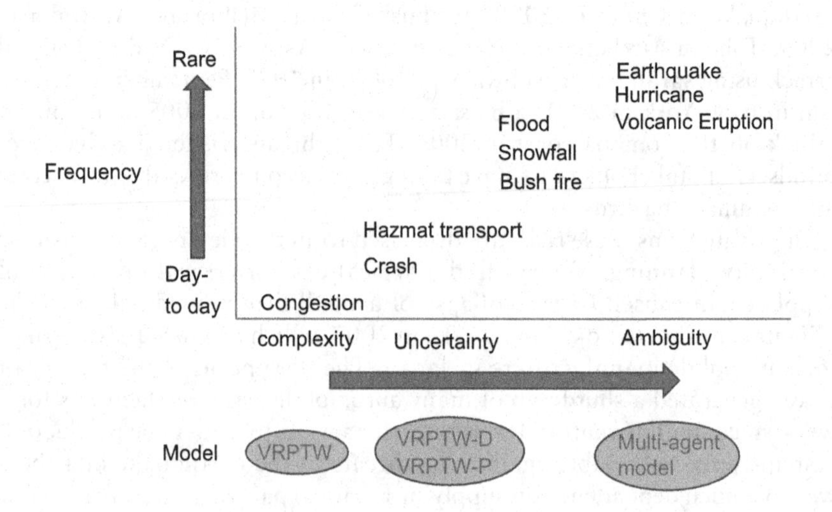

Figure 6.1 Classification of risks related to city logistics.

windows (VRPTW) models can be applied for replicating the movements of pickup/deliver trucks in urban areas. The second level of difficulty comes from uncertainty. Crashes on roadways are not predictable and in particular vehicles carrying hazardous materials often generate huge damage to residents and buildings. In this class of difficulty, we need to develop models taking into account the dynamic and stochastic nature of travel times and connectivity of road networks. Then VRPTW-D (vehicle routing and scheduling with time windows – dynamic) or VRPTW-P (vehicle routing and scheduling with time windows – probabilistic) models are available to replicate the dynamic adaptation of starting time and route choice of pickup/deliver trucks in urban areas based on ITS (Intelligent Transport Systems) applications. The third class of difficulty comes from ambiguity. Bad weather conditions and natural hazards are included in this class and these events occur less often but have a large effect on urban freight transport systems. In this class more sophisticated treatments are required for replicating the behaviour of stakeholders since simply applying VRPTW-D or VRPTW-P models are not sufficient in ambiguous situations due to unpredictable interaction among stakeholders as well as action-reaction relationships between stakeholders and the environment. Multi-objective models or multi-agent models and simulations are effective to assess the effects of these events and evaluate initiatives for responding to them.

Concepts of risk governance are proposed (Kröger, 2008) and can be used to cope with risks of complexity, uncertainty and ambiguity. Risk governance goes beyond risk management and includes the cycle of pre-assessment, appraisal, characterisation and evaluation, and management. The comprehensive framework of risk governance allows us to understand the dependency of each stakeholder related to city logistics initiatives and critical infrastructure as well as the critical points of supply chains.

6.2 TRAVEL TIME RELIABILITY IN CITY LOGISTICS

Incorporating travel time variability in urban goods delivery using road networks is critical for achieving the aims of city logistics for two main reasons: (a) the travel time significantly varies due to heavy congestion on urban roads, and (b) short time windows are typically set for delivery vehicles arriving at customers. To meet the requirements of arriving at the customers with short time windows such as 30 minutes to 1 hour, stochastic vehicle routing and scheduling models with time windows (SVRPTW) where travel times are stochastic can be applied using travel time data provided by ITS and truck GPS.

Numerous researchers have investigated stochastic vehicle routing problems (SVRPs) with uncertain demands of customers (Bertsimas and van Ryzin, 1991; Gendreau et al., 1996; Laporte et al., 2002), whereas only a few researchers have focused on uncertain travel times (Laporte et al., 1992; Lambert et al., 1993; Kenyon and Morton, 2003; Ando and Taniguchi, 2006; Russell and Urban, 2007; Thompson et al., 2011; Ehmke et al., 2015). Since the uncertainty of travel times significantly affects the performance of urban distribution systems, this section describes the stochastic vehicle routing problem with uncertain travel times.

Laporte et al. (1992) studied the vehicle routing problem with stochastic travel times and presented both chance constraint and recourse models. Chance constraint limits the probability of route failure under the certain threshold, whereas the stochastic programme with the recourse penalises the expected value when route travel times exceeds respected targets.

Taniguchi et al. (2001) presented the stochastic vehicle routing and scheduling problem with time windows (SVRPTW) model considering the uncertain travel times. The objective function of their model included the fixed costs, operation costs, early arrival and delay penalty at customers and they assumed soft time windows. They incorporated the penalty of early arrivals and delays. If a vehicle arrives earlier than the starting time of time windows, the freight carrier should wait until the starting time and pay the penalty of early arrival proportional to the waiting time. If the vehicle arrives late, they need to pay the penalty of delay proportional to the delay time. Usually, the rate of delay penalty is higher than the early arrival penalty. They combined the SVRPTW model with traffic simulation which provides the variable travel times and found that the total costs considering the stochastic travel times give 11.3% reduction compared to the case of using fixed mean travel times for a hypothetical road network.

Ando and Taniguchi (2006) investigated the SVRPTW incorporating the uncertainty of travel times based on the travel time estimation using probe vehicle and VICS (Vehicle Information Communication Systems) data. They applied their model in the case of Osaka, Japan and found that the optimal routes of delivery vehicles considering the variable travel times were different from the usual operation of delivery. They showed the routing of usual operation and the optimal solution of the SVRPTW which gives more reliable routing in terms of operation total costs. The authors found that the mean total costs were reduced by 4.1% and the standard deviation decreased by 75.2% in five days of operations.

Ehmke et al. (2015) studied the SVRPTW assuming the travel times are stochastic. They presented a chance-constrained model, where the restrictions are placed on the probability that the individual time windows constraints are violated.

6.3 RISKS IN HAZARDOUS MATERIALS TRANSPORT

Transport of hazardous materials in urban areas has been an important research area for decades. Once a vehicle carrying hazardous material is involved in a crash on a roadway it may cause a large impact on people and buildings and other traffic by explosions or spills of dangerous materials. These risks should be assessed in advance and well controlled for safer management of transport using ICT (Information and Communication Technology) and ITS.

Modelling techniques are required for assessing the risks and evaluating measures to manage hazardous material (hazmat) transport in urban traffic environments. Erkut and Verter (1998) presented an overview of modelling hazardous material transport and pointed out that different risk models suggest different "optimal" paths for a hazmat shipment between a given origin-destination pair. Five categories of risk models are suggested in the literature: (a) Traditional risk, (b) Population exposure, (c) Incident probability, (d) Perceived risk, and (e) Conditional risk.

Incorporating the risks of hazardous material transport often requires multi-objective optimisation models. Giannikos (1998) presented a multi-objective programming model for this problem taking into account total costs, total perceived risk, individual perceived risk and individual disutility. Chang et al. (2005) described a method for finding non-dominated paths for multiple routing objectives in networks where the routing attributes are uncertain, and the probability distributions that describe those attributes vary by time of day. Bell (2006) discussed using a mix of routes by determining the set of safest routes and the safest share of traffic between these routes leads to a better risk averse strategy based on game theory. Beroggi (1994) proposed a real-time routing model for assessing the costs and risks in a real-time environment and pointed out that the ordinal preference model was superior to the utility approach.

Pradahnanga et al. (2014) presented a Pareto-based bi-objective optimisation model of hazardous materials vehicle routing and scheduling problem with time windows (HVRPTW). The model incorporates both the risks of transporting hazmat and the total travel times to serve a given set of customers with time windows. They applied their model in the road network in Osaka with 24 customers. They showed the results of the Pareto front of the bi-objective optimal solutions, which capture a wide range of Pareto-optimal sets. These results also compared with the single objective solution of minimised total travel times and minimised risk value using ant colony techniques. The Pareto fronts were well distributed between these two single optimal solutions.

Szeto et al. (2017) proposed a multi-demon approach to address a hazmat routing and scheduling problem for the general transportation network with multiple-hazmat classes when the incident probability is unknown or inaccurate. They applied the model in Singapore and found that to have the safest shipment of one type of hazmat, different trucks carrying the same type of hazmat need to take different routes and links. In case of multiple-hazmat transportation, it is recommended that different routes and links be used for the shipment of different hazmat types. Moghaddam (2021) developed a multi-objective multi-period optimisation model to find optimal links and routes to maintain a balance between safety and fast distribution of hazmats on the road network. They considered unknown probabilities of hazmat incidents along with game-theoretic demon approach and proposed a solution method using integrated Monte Carlo simulation and fuzzy goal programming to obtain Pareto-optimal solutions.

6.4 RESILIENCE IN DISASTERS

Urban freight transport systems can be disrupted by natural disasters such as earthquakes, floods and volcanic eruptions as well as manmade disasters such as terrorist attacks. However, we need to continue to provide services of goods distribution to customers during the emergency response and recovery stages. City logistics schemes help to adapt to the new situation of the reduced capacity of road networks and logistics facilities caused by disasters. Murray-Tuite (2006) defined ten dimensions of transport resilience: redundancy, diversity, efficiency, autonomous components, strength, collaboration, adaptability, mobility, safety, and the ability to recover quickly.

Several important factors need to be considered in improving resilient urban freight transport systems in disasters: (a) Preparation for disasters, (b) Collaboration among stakeholders in terms of sharing information and resources in disasters, and (c) Mitigation of damages to infrastructures and quick recovery after the event. First, preparation for disaster management is important, including business continuity management (Tómasson, 2022), relocation of distribution centres, and use of alternative transport modes. Second, collaboration among the public and private sectors in terms of sharing information and resources in disasters is necessary.

Chang et al. (2022) presented simulation models of post-disaster multi-commodity distribution under uncertainty. They not only considered commodity distribution from distribution centres to relief centres but also within relief centres to fill the shortage. This model provided a good method for finding the optimal vehicle and inventory routing solutions under varying disaster scenarios when applied in Taiwan.

6.5 CONCLUSIONS

Reliability and resilience should be well addressed in city logistics. Incorporating travel time variability in urban goods delivery using road networks is critical for achieving the aims of city logistics. Stochastic and dynamic vehicle routing and scheduling models can be used to find the optimal solutions for urban delivery with the uncertainty of travel times. The risks of transporting hazardous materials should be assessed in advance and well controlled for safer management. Multi-objective optimisation models are often required for hazmat transport. Important factors for resilient urban freight transport in disasters are: (a) Preparation for disasters, (b) Collaboration among stakeholders in terms of sharing information and resources in disasters, and (c) Mitigation of damages to infrastructures and quick recovery after the event.

REFERENCES

Ando, N. and Taniguchi, E. (2006). Travel time reliability in vehicle routing and scheduling with time windows, *Network Spatial Economics*, **6**, 293–311.

Bell, M.G.H. (2006). Mixed route strategies for the risk-averse shipment of hazardous materials. *Networks and Spatial Economics*, **6**, 253–265.

Beroggi, G.E.G. (1994). A real-time routing model for hazardous materials. *European Journal of Operational Research*, **75**, 508–520.

Bertsimas, D.J. and van Ryzin, G. (1991). Stochastic and dynamic vehicle routing problem in the Euclidean plane, *Operations Research*, **39**, 601–615.

Chang, K.-H, Hsiung, T.-Y., Chang, T.-Y. (2022). Multi-commodity distribution under uncertainty in disaster response phase: Model, solution method, and an empirical study, *European Journal of Operational Research*, **303**, 857–876.

Chang, T.S., Nozick, L.K., & Turnquist, M.A. (2005). Multi-objective path finding in stochastic dynamic networks, with application to routing hazardous materials shipments. *Transportation Science*, **39**, 383–399.

Ehmke, J.F., Campbell, A.M. and Urban, T.L. (2015). Ensuring service levels in routing problems with time windows and stochastic travel times, *European Journal of Operational Research*, **240**, 539–550.

Erkut, E. and Verter, V. (1998). Modeling of transport risk for hazardous materials. *Operations Research*, **46**, 625–642.

Gendreau, M., Laporte, G. and Séguin, R. (1996). A Tabu search heuristic for the vehicle routing problem with stochastic demands and customers. *Operations Research*, **44**, 469–477.

Giannikos, I. (1998). A multi-objective programming model for locating treatment sites and routing hazardous wastes. *European Journal of Operational Research*, **104**, 333–342.

Kenyon, A.S. and Morton, D.P. (2003). Stochastic vehicle routing with random travel times. *Transportation Science*, **37**, 69–82.

Kröger, W. (2008). Critical infrastructures at risk: A need for a new conceptual approach and extended analytical tools. *Reliability Engineering and System Safety*, **93**, 1781–1787.

Lambert, V., Laporte, G. and Louveaux, F. (1993). Designing collection routes through band branches, *Computers and Operations Research*, **20**, 783–791.

Laporte, G., Louveaux, F. and Mercure, H. (1992). The vehicle routing problem with stochastic travel times, *Transportation Science*, **26**, 161–170.

Laporte, G., Louveaux, F.V. and van Hamme, L. (2002). An integer L-shaped algorithm for the capacitated vehicle routing problem with stochastic demands. *Operations Research*, **50**, 415–423.

Moghaddam, K.S. (2021). Multi-objective hazardous materials routing and scheduling for balancing safety and travel time. *International Journal of Applied Management Science*, **13**, 69–93.

Murray-Tuite, P.M. (2006). A comparison of transportation network resilience under simulated system optimum and user equilibrium conditions, *Proceedings of the 2006 Winter Simulation Conference*, Monterey, CA, 1398–1405.

Pradahnanga, R., Taniguchi, E., Yamada, T., and Qureshi, A.G. (2014). Bi-objective decision support system for routing and scheduling of hazardous materials, *Socio-Economic Planning Sciences*, **48**, 135–148.

Russell, R.A. and Urban, T.L. (2008). Vehicle routing with soft time windows and Erlang travel times, *Journal of the Operational Research Society*, **59**, 1220–1228.

Szeto, W.Y., Farahani, R.Z., Sumalee, A. (2017). Link-based multi-class hazmat routing-scheduling problem: A multiple demon approach, *European Journal of Operational Research*, **261**, 337–354.

Taniguchi, E., Thompson, R.G., Yamada, T. and van Duin, R. (2001). *City logistics – Network Modelling and Intelligent Transport Systems*. Elsevier, Pergamon, Oxford.

Thompson, R.G., Taniguchi, E. and Yamada, T. (2011). Estimating benefits of considering travel time variability in urban distribution, *Transportation Research Record*, **2238**, 86–96.

Tómasson, B. (2022). Using business continuity methodology for improving national disaster risk management, *Journal of Contingencies and Crisis Management*, **31**, 1–15.

Chapter 7

Evaluation

7.1 INTRODUCTION

Evaluation is an appraisal procedure that links the predicted consequences of alternatives to the selection of an appropriate course of action. It is a key element of the systems approach to City Logistics (Taniguchi et al., 2001). There is a need for models and software to provide information to assist decision makers improve the sustainability of urban freight systems. Since data only becomes information when it is communicated in forms and times which are suitable for use in a decision making environment, analytical procedures are required to transform data generated from models into information by processing and structuring output to make it more understandable and relevant to problems. Data analytics methods are needed to enhance output from models to make them more comprehensible and meaningful.

Evaluation procedures should assist the decision maker in reaching informed decisions. They should provide information to decision makers and interested parties in a comparable form, highlighting the trade-offs from estimates of the consequences (impacts) of new initiatives or policies.

Evaluation is the process of methodically comparing alternative courses of action, with the aim to determine the best course of action. Urban freight systems have multiple objectives and criteria with a number of options that can address problems (Thompson and Hassall, 2006; Thompson, 2015). There is a need for models and software to provide useful information to examine the merit of options to assist decision makers determine the best options.

Urban freight stakeholders typically have different objectives and performance measures. Shippers, carriers and receivers are generally interested in minimising the financial costs of logistics well as improving reliability and levels of service (Table 7.1).

Analytics should be developed to highlight the trade-offs in the predicted performance of options and illustrate how sensitive these are to key

DOI: 10.1201/9781003261513-8

Table 7.1 Typical performance measures

Shippers	*Carriers*	*Receivers*
Receiver Satisfaction	Receiver Satisfaction	Delivery Punctuality
Punctual and Secure Pickups with No Damage	Shipper Satisfaction	Delivery Security
Transportation Costs	Transport Costs	Convenient Deliveries Times

Source: Adapted from Macharis et al. (2016); Perera & Thompson (2021); Aljohani & Thompson (2018).

assumptions within models. There is a need to provide output that incorporates uncertainty in the future performance of options.

This chapter presents a range of analytics methods that can be used for evaluating options to improve the sustainability of urban freight systems. Financial analysis, multi-criteria assessment, and multi-objective optimisation problems are covered. Applications are described for a range of City Logistics initiatives, including collaborative freight networks, electric freight vehicles, urban consolidation centres, and road tolls.

7.2 NETWORKS

When considering the overall performance of urban freight systems, it is crucial to consider the efficiency of networks. This can be measured by ratio of amount and distance of goods moved to the distance travelled by freight vehicles (ITF, 2018). In terms of weight, network efficiency (NE) can be expressed as TKM/VKT, where TKM is the tonne kilometres moved and VKT is vehicle kilometres travelled. Improving the efficiency of urban freight networks for a fixed level of freight demand involves reducing the total distance travelled by vehicles that requires increasing the utilisation of the capacity of vehicles. It is therefore important for urban freight analytics be developed for evaluating the effectiveness of initiatives aimed at improving network efficiency. This will entail monitoring and analysing empty or unladen trips and vehicle load factors.

Network simulation models can be developed to predict changes to the physical productivity of urban freight networks using Performance Based Standards (PBS) vehicles (Hassall and Thompson, 2012; Thompson and Hassall, 2014). PBS vehicles that have higher capacity can substantially increase productivity. Metrics for investigating productivity include total kilometres travelled, total operational hours and individual fleet numbers.

Route stem time is the time from when a vehicle leaves the warehouse to the time when it makes its initial delivery. It is related to the distance from warehouses to where customers are located. In large metropolitan areas stem times for home deliveries from e-Commerce can be considerable. This can lead to low levels of productivity and high transport costs.

Urban distribution routes typically involve dropping off goods at several customers. A vehicle's load factor measures the utilisation of a vehicle's load-carrying capacity in terms of weight, LF(w) or volume LF(v) over a vehicles route. Load factors can be calculated for numerous vehicles operating over a distribution network.

Optimisation procedures used for designing urban distribution networks generally aim to minimise the number of vehicles (NVs) used as well as the total distance travelled (VKT). The radar plot is an effective way of displaying the performance of alternatives for a number of criteria. This type of plot can be produced in software such as Excel, MatLab, and R.

7.2.1 Retail swap networks

A common problem in retailing is swapping goods between stores within the same urban area. A simple problem can be used to illustrate the performance of a range of different network configurations, where five stores need to exchange 200 kg between all other stores using vans with a capacity of 1,200 kg. A number of network options are available (Figure 7.1). The

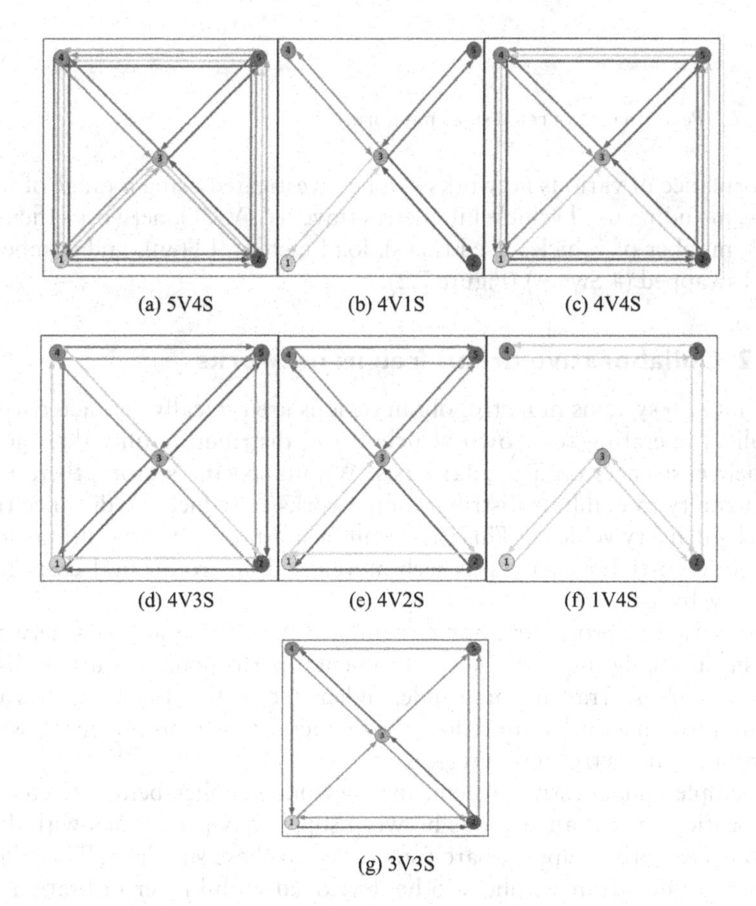

Figure 7.1 Retail swap networks.

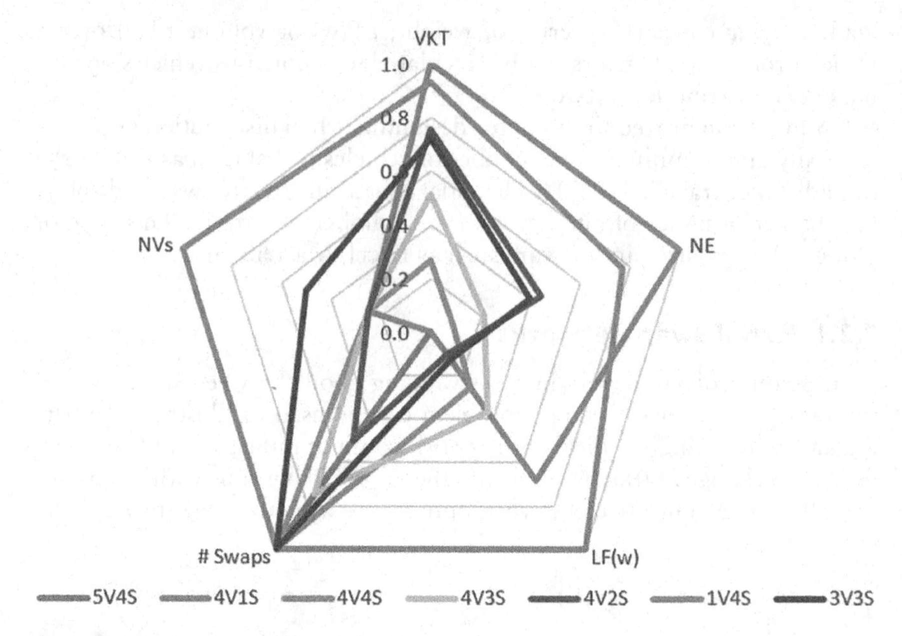

Figure 7.2 Performance of retail swap networks.

performance of various networks can be investigated using a range of measures, including total vehicle kilometres travelled (VKT), network efficiency (NE), number of vehicles used (NVs), load factors (LF(w)), and number of loads swapped (# Swaps) (Figure 7.2).

7.2.2 Collaborative urban freight networks

Distribution systems in metropolitan regions are typically characterised by suppliers operating their own vehicle fleets, distributing only their goods to their customers on a regular basis. Within specific sectors, there is an opportunity to combine distribution networks to reduce the distance travelled by delivery vehicles. This can result in substantial savings in transport operating costs for carriers as well as reduced environmental costs from freight vehicles.

Optimisation procedures for designing collaborative logistics networks can substantially improve the environmental performance of urban distribution systems. Transforming independent retail distribution chains into collaborative networks can reduce the financial and environmental costs of distribution in metropolitan regions.

A simple collaborative system, involves one supplier being selected for the location to exchange goods between suppliers where goods with destinations near other suppliers are transferred to these suppliers. This allows delivery routes from suppliers to be developed with higher utilisation and substantially lower travel distances.

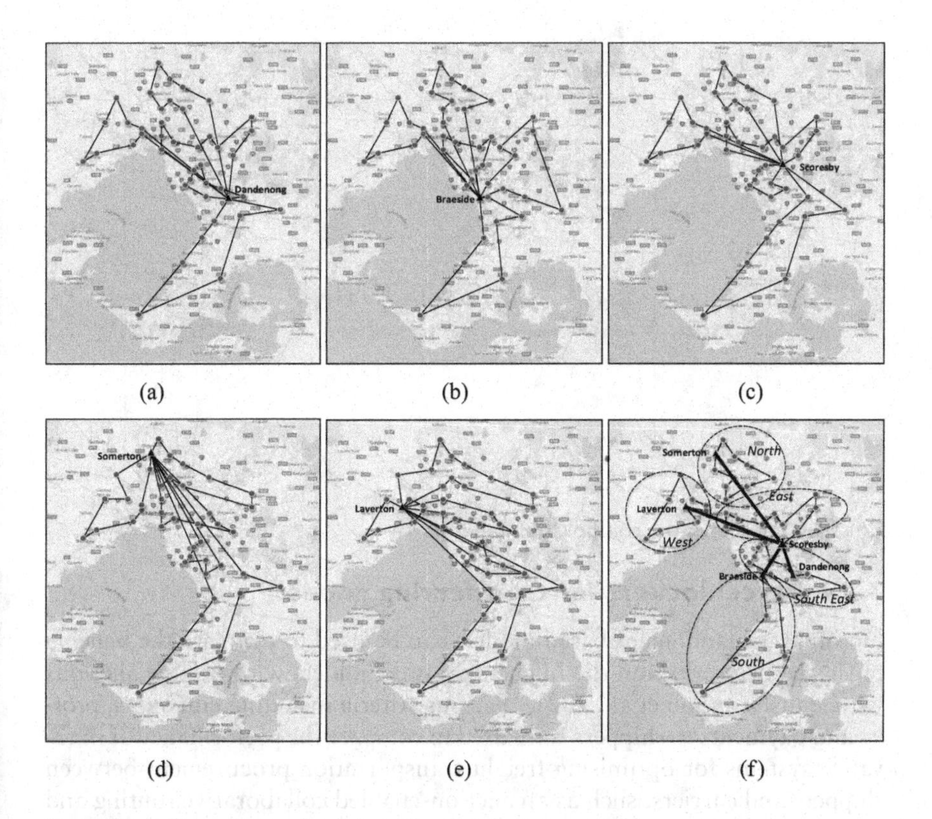

Figure 7.3 Independent and collaborative distribution networks.

Independent distribution networks (Figure 7.3a–e) can be transformed into a collaborative network (Figure 7.3f) for distributing goods to a large hardware retailer in Melbourne. A substantial reduction in the distance travelled by vans and stem times as well as increased load factors and overall network efficiency were estimated for a retail distribution network in Melbourne using existing warehouses and vehicles with no additional vehicles or warehouses required (Figure 7.4).

7.2.3 Urban shuttle

Shared freight networks can be designed to reduce empty return truck trips travelling back to warehouses in large metropolitan areas. Increased network efficiency can be achieved which has the potential to substantially reduce urban traffic congestion. MatLab was used to develop a network cost model to compare the performance of independent and shared freight networks with PBS vehicles in metropolitan Melbourne (Thompson et al., 2020). The complete shared network with PBS vehicles was estimated to reduce the total distance travelled by approximately 85% and increase network efficiency from 2.5 to 17.2 pallets/veh.

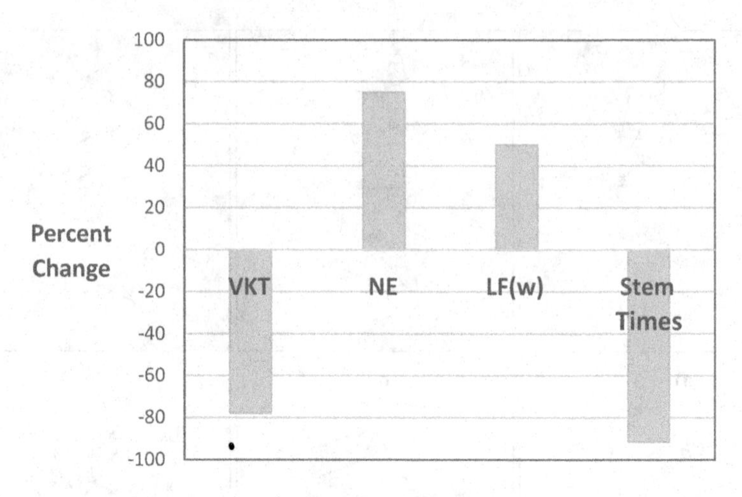

Figure 7.4 Impacts of the collaborative network.

7.2.4 Parcel lockers for transferring goods

Optimisation and agent based models can be used to estimate the benefits of shared networks and the impacts on stakeholders with on-line auctions with transfers (Guo et al., 2021). Various criteria including emissions, profits and payments to shippers were used to compare the performance of innovative systems for optimising freight transportation procurement between shippers and carriers, such as an auction-enabled collaborative routing and an auction-enabled collaborative routing with transfers.

7.3 FINANCIAL ANALYSIS

Analytics can be applied to assess the financial viability of initiatives such as electric freight vehicles (EFVs) and urban consolidation centres (UCCs). Carriers need to investigate options relating to alternative fuel vehicles for improving the sustainability of urban freight operations. Governments need to understand how consolidation centres can make a profit.

7.3.1 Electric freight vehicles

There are numerous financial costs incurred by fleet owners associated with owning and operating freight vehicles including capital, energy, maintenance and taxation. These are largely influenced by the size of the fleet and the distance travelled by vehicles.

Since there is considerable variation in the initial costs of batteries and their life as well as charging systems and resale value, lifecycle cost analysis as well as sensitivity should be used to compare costs for electric freight vehicles and diesel vehicles.

Lifecycle cost analysis can be undertaken by estimating the Net Present Value (NPV) or Present Worth (PW) that allows initial, ongoing and end-of-life financial flows to be considered for comparing various types of freight vehicles. Financial models can be developed for different vehicles based on the nature of freight networks operated that can influence the distances vehicles travel. Changes in NPV for various scenarios and changes from expected operating costs provide useful information.

The effects of fleet size, network configuration and demand densities on NPV can be investigated using financial modelling. The relative contribution of different costs across charging scenarios can be estimated. The effects of general cost factors such as discount rate, vehicle utilisation, emissions costs and planning horizon as well as vehicle cost factors such as capital cost, depreciation rate, fuel price, fuel inflation rate, fuel efficiency and maintenance costs on overall financial performance can be estimated. Forecasted estimates of costs are used to compare alternatives.

The overall financial costs of electric and diesel freight vehicles can be compared by subtracting the NPV of the costs of electric vehicles, NPV(EV) from the NPV costs of diesel vehicles, NPV(DV).

Using the parameters listed in Table 7.2, the NPV of an electric freight vehicle compared with a diesel was estimated to be $-5,420. This indicates the expected costs of an electric truck over the planning horizon are higher than those for a diesel vehicle.

However, due to the uncertainty of the future values of costs, it is important to undertake a sensitivity analysis. Spider plots and tornado charts provide an effective means of illustrating the magnitude of changes to NPV for changes to individual parameters (Eschenbach, 1992).

Table 7.2 Vehicle cost parameters

		Electric vehicle (EV)	*Diesel vehicle (DV)*
Discount Rate (%)	4		
Planning Horizon (yrs)	5		
Utilisation (km/yr)	80,000		
Capital Cost ($)		90,000	50,000
Salvage Value (%)		30	30
Maintenance Cost ($/km)		0.01	0.02
Electricity Price ($ kWh)		0.2	
EV Efficiency (KW h/km)		0.5	
CO_2 Emissions Cost ($/ton)			20
Diesel Price ($/l)			1.5
Diesel Price Inflation (%)			0
DV Fuel Efficiency (km/L)			10
Emissions (ton/km)			0.00047

Spider plots provide a useful means of identifying parameters that can have a large influence on the viability of alternatives. The effects of changes in cost parameters from base levels on NPV can be compared. Figure 7.5 presents a spider plot for comparing the financial performance of an electric truck with a diesel truck. This plot highlights the major effect changes in the price of the EV, electricity price and diesel price would have on the financial viability of the EV. The relatively small effect of CO_2 emissions costs can also be observed.

Tornado diagrams aim to identify parameters that are likely to have a major influence on the performance of an option by ranking parameters by the range of variation from nominated lower and upper limits. A tornado diagram for evaluating the electric vehicle using the data presented in Table 7.3 is presented in Figure 7.6. This shows that variation in the price of diesel and electricity within the specified ranges will have the most effect on NPV.

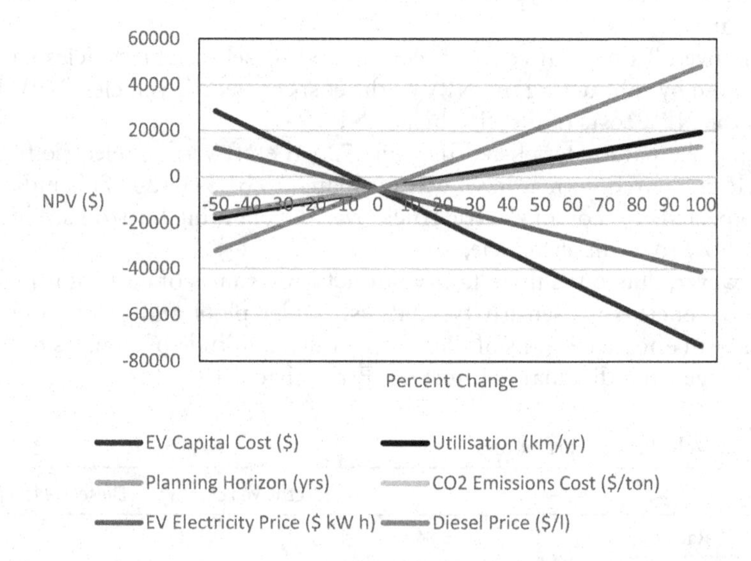

Figure 7.5 Spider plot.

Table 7.3 Ranges of Parameters

	Lower limit	*Base case*	*Upper limit*
EV Capital Cost ($)	75,000	90,000	95,000
Utilisation (km/yr)	60,000	80,000	85,000
Planning Horizon (yrs)	4	5	8
CO_2 Emissions Cost ($/ton)	18	20	40
EV Electricity Price ($ kWh)	0.1	0.2	0.4
Diesel Price ($/l)	1.4	1.5	3

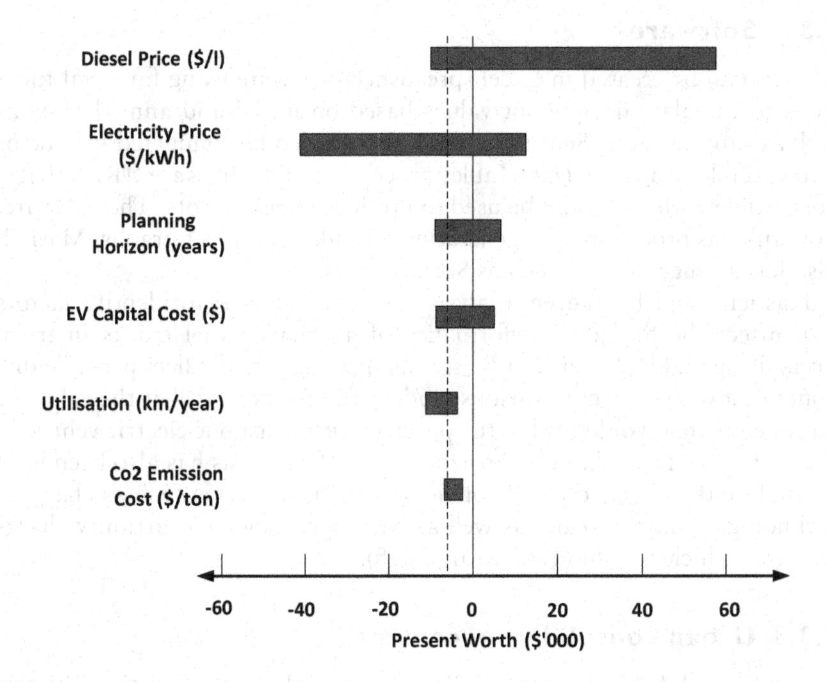

Figure 7.6 Tornado diagram.

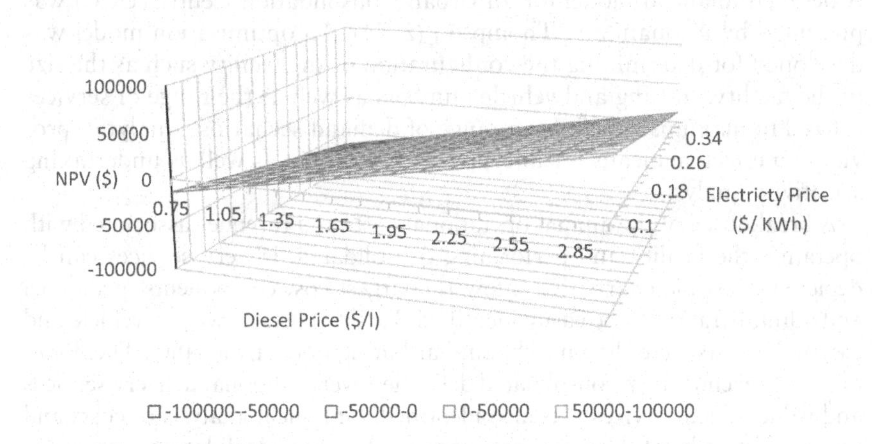

Figure 7.7 Surface chart using a two-way data-table in Excel.

It is possible to investigate the effects of variation of two parameters on the financial viability of options. Figure 7.7 presents a surface plot of the EV evaluation produced using the data-table and 3D surface plot in Microsoft Excel. This plot shows the improved viability for the EV that occurs with both higher diesel prices as well lower electricity prices.

7.3.2 Software

Models can be created in Excel spreadsheet software using financial functions to calculate net present values based on initial and annual costs as well as salvage value. Sensitivity analysis can also be conducted for one or two variables using the Data Tables procedures. The software @Risk that is compatible with Excel can be used to produce tornado plots. The software, Toprank has procedures for generating tornado and spider graphs. MatLab also has a range of plots such as Surface Plots.

Elasticity and breakeven analysis can be conducted to identify factors that affect the financial performance of alternative fuel trucks in truck fleets. Feng and Figliozzi (2013) estimated changes to the fleet per-mile discount cost with respect to various model parameters as well as the values of parameters that would lead to the purchase of at least one electric vehicle in the first year of the planning horizon. Lifecycle analysis has also been used to analyse the effects on NPV of electric truck scenarios such as charging technologies and strategies as well as battery capacity, opportunity charging, and vehicle weight (Teoh et al., 2018).

7.3.3 Urban consolidation centres

Urban Consolidation Centres (UCCs) are a popular City Logistics initiative that has good potential for improving sustainability of urban distribution. A detailed financial model for an Urban Consolidation Centre (UCC) was presented by Aljohani and Thompson (2021). An optimisation model was developed for determining the configuration of the facility such as the size of the facility, staffing and vehicle numbers as well as the range of services offered to maximise profit for a range of demand scenarios. Analytics provides an effective means for interpreting key output as well as undertaking sensitivity analysis.

A breakdown of the annual operational costs and revenues associated with operating the facility and performing consolidated delivery services can be depicted using pie charts that allow the largest cost components, personnel and administration to be easily identified. Facility costs as well as vehicle and cargobike costs were shown to be similar but significantly smaller. The dominance of revenue from consolidated deliveries over additional delivery services and value-added services was also identified. Graphical analysis of costs and revenue can be helpful for planners of urban freight consolidation centres.

Sensitivity analysis allowed the effect of operating different classes of delivery vans on the facility's profitability to be examined. The model assumes that large LCVs would be used. Smaller vehicles with less capacity would require more vehicles and drivers decreasing profit, whilst extra-large vehicles would substantially increase profit.

Radar plots were used to undertake sensitivity analysis to investigate the effect on net profit margins of four operational variables across three demand scenarios to assist operators identify factors that can achieve gains in profitability. With the lower-limit scenario, only one factor, increasing additional services comes close to achieving a profit. This allowed vehicle load utilisation levels to be identified as the factor for achieving the largest increase in profit.

7.4 MULTI-CRITERIA ANALYSIS (MCA)

It is common for multiple criteria to be considered to compare the performance of options. This mainly arises due to the vast range of factors relating to sustainability and efficiency as well as the numerous stakeholders involved in urban freight systems.

Although the general aim of multi-criteria analysis is to determine a ranking of alternatives, analytics can provide an important role in highlighting the differences in performance between alternatives for various performance criteria and identifying trade-offs between alternatives. A review of criteria and multi-criteria decision making methods is presented by Jamshidi et al. (2019).

This section presents several plots and charts that can provide an effective means of illustrating the differences in performance of alternatives across a range of factors.

7.4.1 Radar plot

The radar plot can be used to display the performance of alternatives for various criteria. Figure 7.8 shows the attributes and performance of several electric vans and trucks using the extreme value standardisation method (Aljohani and Thompson, 2018). It can be observed that the Renault Master ZE vehicle performs the best for three criteria (Volume Capacity, Weight Capacity, and Battery Charing Time), whilst the Peugeot Partner Electric and Nissan e-NV200 vehicles performs the best for one criteria (Price).

7.4.2 Priority weightings

Many multi-criteria analysis methods require the estimation of importance weights for criteria. The Analytical Hierarchy Process (AHP) provides procedures for grouping criteria as well as estimating criteria weightings based on pairwise judgements. An effective means of displaying such estimated weightings is a Tree map that was used to show the relative importance of weightings estimated for determining the siting of a UCC using fuzzy AHP (Aljohani and Thompson, 2020). This allows the dominance of the suitability criteria over transport accessibility and logistics land-use attributes criteria to be shown. Scenarios can be used to represent variations to the

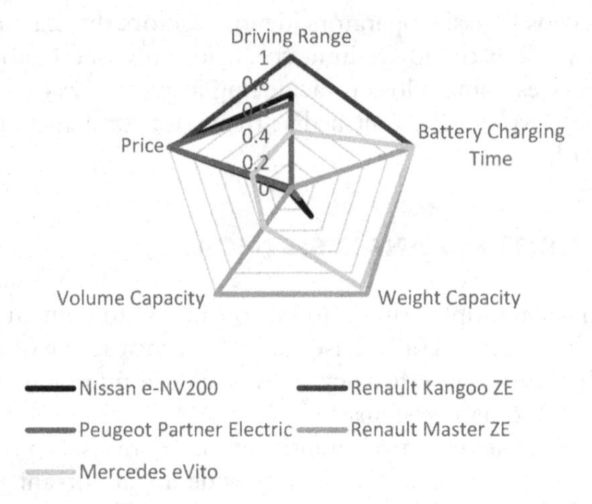

Figure 7.8 Performance of electric vans and trucks.

estimated weightings to investigate how sensitive the rankings of sites are to these. The variability in ranking between scenarios can be highlighted.

7.4.3 Promethee

Promethee is a popular multi-criteria method for comparing alternatives. Geometrical Analysis for Interactive Aid (GAIA) provides a useful tool for interpreting the output from an evaluation using Promethee. An evaluation of 36 electric vehicles for City Logistics, considering performance (Carrying Capacity, Max Velocity, Travel Range), engine (Power and Torque), battery (Charging Time 100% & 80%, Capacity) and Price was undertaken by Watróbski et al. (2017).

GAIA showed that the Battery Capacity criterion conflicts with the Battery Charging Time criteria with the greater the Battery Capacity, the longer its charging time. In contrast, the pairs of Carrying Capacity and Battery Capacity, as well as Engine torque and Engine power express similar preferences, *i.e.*, the value of Engine torque increases along with the Engine power value and the Carrying Capacity grows together with the Battery Capacity. The Travel Range criterion was shown to be unrelated to the Battery Capacity and Carrying Capacity in terms of preferences. High-ranking alternatives A29 (Ranger EV Ford), A9 (e-Wolf Omega 0.7) and A8 (e-NV200+ Nissan) were shown to be most supported by the Max Velocity and Travel Range criteria, whilst alternatives A15 (Mitsubishi Minicab-MiEV), A16 (Mitsubishi Minicab-MiEV, 10.5 kWh), and A17 (Mitsubishi Minicab-MiEV, 16 kWh) are supported by the Price and Battery Charging Time criteria.

Similar analysis conducted with criteria groupings indicated that the Price criteria group is in slight conflict with the Engine group, whilst the Engine group is not related to the Battery group, as well as the

Performance group values grow independently from the Price group in terms of preference. The Engine group supports alternatives A6 (EVI MD) and A7 (EVI's Walk-in van), whereas the Price group has a positive effect on the ranking position of options A8, A15–A17 and A23 (Peugeot's Partner Panel Van).

Sensitivity analysis can be used to investigate how variation in the importance weightings of clusters of criteria can affect overall vehicle rankings. The ranges of stability of the Performance and Battery clusters were shown to be much wider than the ranges of stability of the Engine and Price clusters. If the criteria from the Performance cluster were assigned a much higher weight (over 66%), A27 (Phoenix Motorcars SUV) alternative would become the top ranking alternative. Furthermore, if the weight of the Engine cluster was above 43%, the A6 (EVI MD) alternative would become the leading one. Alternatively, if the Engine cluster was considered less important and its weight dropped below 5%, the alternative A8 (Nissan e-NV200+) would be the top of the ranking vehicle.

Subsequently, the various weights possibly assigned to each of the criteria were analysed in a sensitivity analysis. The effects of individual importance criteria weightings criteria on overall rankings of vehicles can be analysed using ranking plots. It was shown that A7 (Electric Vehicles International, Walk-in Van) climbs 16 slots (from 20th to 4th ranking) when the Battery Charging Time (100%) criteria increases from very low to very high.

7.4.4 Combined methods

Tadić et al. (2018) defined a set of social, economic and ecological criteria for assessing City Logistics schemes. Social criteria consist of mobility, attractiveness of the city, freeing up public space, traffic accidents as well as noise and vibration. Economic criteria consist of reliability and accuracy of delivery, costs of implementation and control as well as accessibility. Ecological criteria considered were energy savings and air pollution. Initiatives evaluated were toll collection, restrictions on the vehicle capacity and size, vehicle loading factor control, low-emission zones, access time windows, night delivery, infrastructure reservation as well as loading and unloading zones. A hybrid MCDM model that combines Delphi, AHP, and SWARA methods in a fuzzy environment is presented.

Recently there is a growing interest in assessing the maturity of cities concerning planning and implementing City Logistics solutions. A classification scheme has been developed for identifying a cities maturity level based on various criteria that measure the degree of advancement in both the formulation and implementation of strategies (Kiba-Janiak et al., 2021). Several groups of criteria have been defined relating to solutions implemented, including infrastructure, use and management of land, conditions of access to selected zones of the city, managing freight traffic, promotion of ecological freight transport, standards (regulations) of collecting information and tasks realised by local government in the area of sustainable integrated passenger and freight transport (ecological, social, and economic).

7.4.5 Knowledge management

Knowledge management tools can assist with identifying potential City Logistics solutions. The "Initiative Selector" has been developed to provide information regarding possible alternatives for various problems (https://cite.rpi.edu/iselector). To aid the search for options, details of the nature of the problem, geographic scope and source of the problem can be specified (Holguin-Veras et al., 2021). Details of stakeholders, advantages and disadvantages, examples as well as references are provided for initiatives.

An urban freight implementation tool has been developed using a strategic resource matrix based on a review of strategic options can be used to identify strategies for addressing specific problems (TRB, 2018). Users can select criteria including strategic type, barriers, spatial scope and effectiveness. Recommended strategies and fact sheets are provided.

7.5 MULTI-ACTOR MULTI-CRITERIA ANALYSIS

It is vital for the successful implementation of City Logistics solutions that acceptance from a broad range of stakeholders be achieved. Multi-criteria analysis methods have been adapted to incorporate multiple stakeholders. In particular, the Multi-Actor Multi-Criteria Analysis (MAMCA) methodology provides a useful means of evaluating schemes taking into consideration the acceptance of various stakeholders (Macharis et al., 2009).

Generating a ranking of alternatives using MAMCA involves a number of steps including determining stakeholder groups, identifying alternatives, determine objectives and criteria for each stakeholder group, estimating the importance of criteria for each stakeholder group and evaluating the performance of alternatives (for each stakeholder group's criteria). MAMCA has been used to evaluate numerous urban freight projects including, construction logistics (Macharis et al., 2016), configurations for delivery fleets for consolidated delivery (Aljohani and Thompson, 2018) and tolls for trucks on urban freeways (Perera and Thompson, 2021).

This section illustrates how the MAMCA approach can be used to evaluate options related to urban toll schemes. A number of alternative solutions for determining toll levels for urban freight vehicles were investigated including, minimising environmental costs, minimising social costs, optimising multiple objectives and maximising toll revenue (Perera and Thompson, 2021). Stakeholders considered were freight carriers, government, residents and toll road operators. The evaluation was conducted using the MAMCA software (MAMCA decision making tool).

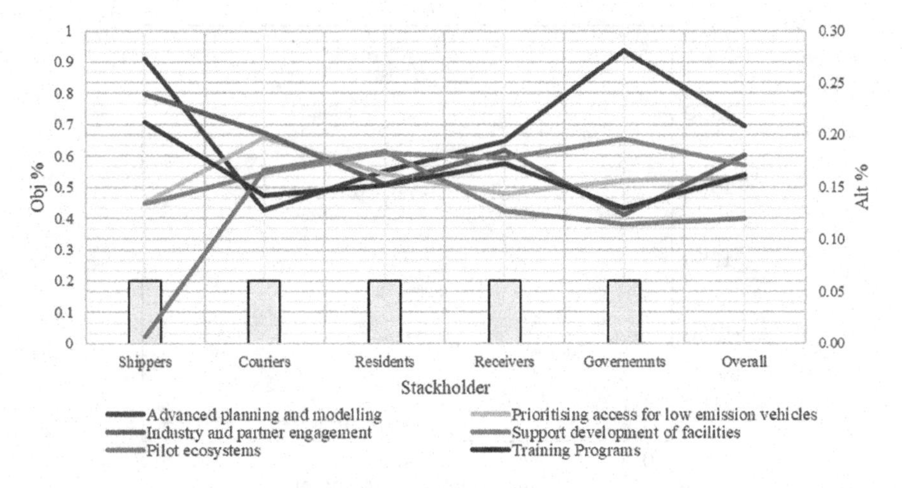

Figure 7.9 MAMCA summary.

MAMCA was recently used to identify government programmes for addressing the impacts of COVID-19 on distribution networks in Sydney. Surveys of government, carriers, shippers, residents and receivers were conducted to evaluate a range of potential programmes based on the assessment of how each programme would satisfy the objectives of each stakeholder (Mohri et al., 2022). Advanced planning and modelling were found to perform well for a majority of stakeholders (Figure 7.9).

7.6 MULTI-OBJECTIVE OPTIMISATION

It is common when modelling urban freight systems for there to be more than one objective. This can arise from trying to optimise several stakeholders' objectives or when transport costs such as time, emissions or operating costs need to be combined and minimised.

Multi-objective optimisation methods utilising mathematical programming and network analysis can be used to determine feasible and optimal solutions for a variety of urban freight problems. Multi-objective optimisation techniques can be used to identify solutions that are superior to the rest of the feasible solutions considering multiple objectives without specifying objective weightings. Pareto optimal solutions are solutions not dominated by any other feasible solution where a better value of any objective is possible only with a poorer value of at least one other objective.

Numerous procedures based on genetic algorithms have been developed for identifying Pareto optimal solutions, including the Vector-Evaluated Genetic Algorithm (Schaffer, 1985), the Non-dominated Sorting Genetic Algorithm (Srinivas and Deb, 1994), the NSGA-II (Deb et al., 2002), the Normalised Normal Constraint Method (Messac et al.,

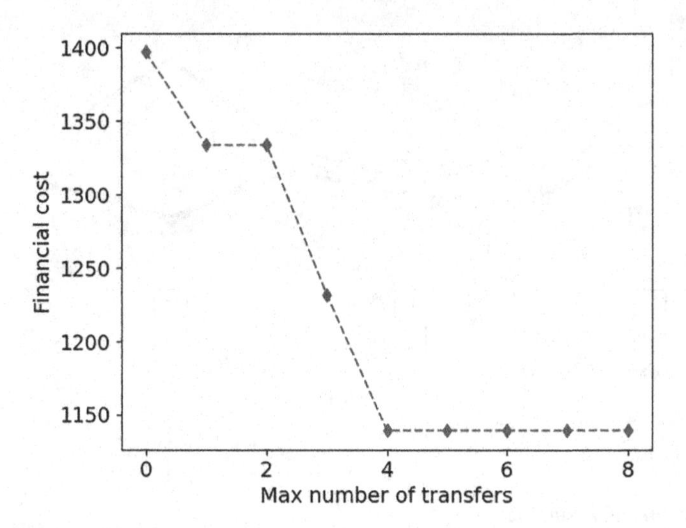

Figure 7.10 Pareto Front for retail swap network.

2003), and the bi-level multi-objective Taguchi genetic algorithm (Xiong and Wang, 2014).

Graphs provide a useful means for depicting solutions on the Pareto front that represents the set of Pareto optimal solutions. Figure 7.10 shows the Pareto front for the retail swap network design problem illustrating the trade-off between financial cost and the maximum number of transfers.

7.7 CONCLUSIONS

Evaluation and sensitivity analysis are important elements in the systems approach to City Logistics. Analytics provides an effective means for investigating uncertainty and conducting what-if analysis.

There is an increasing need for improved analytics to assist decision makers redesign distribution networks, select cleaner freight vehicles, and determine road pricing options for developing more efficient and sustainable urban freight systems.

Urban freight analytics needs to incorporate multiple criteria relating to the performance of alternatives to highlight the trade-offs between options based on financial, social and environmental measures. Analysis procedures also need to illustrate the effects of options for multiple stakeholders with different objectives. This chapter describes methods that allow multiple-criteria and key stakeholders objectives to be considered for evaluating urban freight initiatives.

REFERENCES

Aljohani, K. and Thompson, R.G. (2018). A stakeholder-based evaluation of the most suitable and sustainable delivery fleet for freight consolidation policies in the inner-city area, *Sustainability*, **11**, 124.

Aljohani, K. and Thompson, R.G. (2020). A multi-criteria spatial evaluation framework to optimise the siting of freight consolidation facilities in inner-city areas, *Transportation Research Part A*, **138**, 51–69.

Aljohani, K. and Thompson, R.G. (2021). Profitability of freight consolidation facilities: A detailed cost analysis based on theoretical modelling, Research in Transportation Economics, **90**, 101122, https://doi.org/10.1016/j.retrec.2021.101122.

Deb, K., Pratap, A., Agarwal, S. and Meyarivan, T. (2002). A fast and elitist multiobjective genetic algorithm: NSGA-II. *IEEE Transactions on Evolutionary Computation*, **6**, 182–197.

Eschenbach, T.G. (1992). Spiderplots versus tornado diagrams for sensitivity analysis, *Interfaces*, **22**, 40–46.

Guo, C., Thompson, R.G., Foliente, G. and Kong, X.T.R. (2021). An auction-enabled collaborative routing mechanism for omnichannel on-demand logistics through transshipment, *Transportation Research Part E*, **146**, 102206.

Feng, W. and Figliozzi, M. (2013). An economic and technological analysis of the key factors affecting the competitiveness of electric commercial vehicles: A case study from the USA market. Transportation Research Part C, 26, 135–145.

Hassall, K. and Thompson, R.G. (2012). Determining productivity gains from performance-based standards vehicles in urban areas, Proceedings *91st Transportation Research Board Annual Meeting*, Washington, DC, January 2012, 22–26, 12–1887).

Holguin-Veras J., Ramirez-Rios D., Ng J., Wojtowicz J., Haake D., Lawson C.T., Calderón O., Caron B. and Wang, C. (2021). Freight-efficient land uses: Methodology, strategies, and tools. *Sustainability*, **13**, 3059.

ITF (2018). *Towards road freight decarbonisation – Trends, measures and policies*, International Transport Forum, Paris.

Jamshidi, A., Jamshidi, F., Ait-Kadi, D. and Ramudhin, A. (2019). A review of priority criteria and decision-making methods applied in selection of sustainable city logistics initiatives and collaboration partners, *International Journal of Production Research*, **57**, 15–16, 5175–5193.

Kiba-Janiak, M., Thompson, R.G. and Cheba, K. (2021). An assessment tool of the formulation and implementation a sustainable integrated passenger and freight transport strategies. An example of selected European and Australian cities, *Sustainable Cities and Society*, **71**, 102966.

Macharis, C., Kin, B., Balm, S. and van Amstel, W.P. (2016). Multiactor participatory decision making in urban construction logistics, *Transportation Research Record: Journal of the Transportation Research Board*, **16**, 83–90.

Macharis, C., de Witte, A. and Ampe, J. (2009). The multi-actor, multi-criteria analysis methodology (MAMCA) for the evaluation of transport projects: Theory and practice. *Journal of Advanced Transportation*, **43**, 183–202.

Messac A., Ismail-Yahaya, A. and Mattson, C.A. (2003). Normalized normal constraint method for generating the Pareto frontier, *Structural Multidisciplinary Optimisation*, **25**, 96–98.

Mohri, S., Vijay, A., Kahalimoghadam, M., Stokoe, M., Nassir, N. and Thompson, R.G. (2022). Evaluating initiatives for improving urban freight deliveries: A case study of Sydney, Proceedings, *Australian Transport Research Forum (ATRF)*, Adelaide, Australia, September 28–30, 2022.

Perera, L. and Thompson, R.G. (2021). Multi-stakeholder acceptance of optimum toll schemes, *Research in Transportation Business and Management*, **41**, 100654.

Schaffer, J.D. (1985). Multiple objective optimization with vector evaluated genetic algorithms, In: Grefenstette, J.J. (eds.), *Proceedings 1st International Conference on Genetic Algorithms and their Applications*, Carnegie-Mellon University, July 24–26, 1985.

Srinivas, N. and Deb, K., (1994). Multi-objective optimization using non-dominated sorting in genetic algorithms, *Evolutionary Computation*, **2**, 221–248.

Tadić S., S. Zečević, M. Krstić (2018). Assessment of the political city logistics initiatives sustainability, *Transportation Research Procedia*, **30**, 285–294.

Taniguchi, E., Thompson, R.G., Yamada, T. and Van Duin, R. (2001). *City Logistics – Network Modelling and Intelligent Transport Systems*, Elsevier, Pergamon, Oxford.

Teoh, T., Kunze, O., Teo, C. and Wong, Y.D. (2018). Decarbonisation of urban freight transport using electric vehicles and opportunity charging, *Sustainability*, **10**, 3258.

Thompson, R.G. (2015). Evaluating city logistics schemes, In: Taniguchi, E. and Thompson, R.G. (eds.), City Logistics: *Mapping the Future*, CRC Press, Boca Raton, 101–114.

Thompson, R.G. and Hassall, K. (2006). A methodology for evaluating urban freight projects, In: Taniguchi, E. and Thompson, R.G. (eds.), *Recent Advances in City Logistics*, Elsevier, Amsterdam, 283–292.

Thompson, R.G. and Hassall, K. (2014). Implementing high productivity freight vehicles in urban areas, *Procedia - Social and Behavioral Sciences*, **151**, 318–332.

Thompson, R.G., Nassir, N. and Frauenfelder, P. (2020). Shared freight networks in metropolitan areas, *Transportation Research Procedia*, **46**, 204–211.

TRB (2018). *Tools to Facilitate Implementation of Effective Metropolitan Freight Transportation Strategies*, NCHRP 897 Research Report, Washington DC.

Watróbski, J., Małecki K., K. Kijewska, S. Iwan, A. Karczmarczyk and R.G. Thompson (2017). Multi-criteria analysis of electric vans for city logistics, *Sustainability*, **9**, 1453.Xiong, G. and Wang, Y. (2014). Best routes selection in multimodal networks using multi-objective genetic algorithm. *Journal of Combinatorial Optimization*, **28**, 655–673.

Part 2

Applications

Chapter 8

Autonomous vehicles and robots

8.1 INTRODUCTION

Last-mile delivery is the most critical and most challenging part of urban freight. With the increasing number and ease of online shopping facilities, the demand for last-mile delivery is increasing at a rapid pace. Growing competition among the online shopping platforms has led to many innovations in this business stream including the concepts of fast, free and time-scheduled deliveries to customers. This situation has put further pressure on freight carriers involved in the last-mile deliveries, as they already have to cope with issues such as traffic congestion, lack of parking spaces, re-deliveries and lack of human resources. Recently due to the COVID-19 pandemic, deliveries to homes via e-Commerce have further increased in many cities creating new challenges such as contactless deliveries. To maintain their economic viability as well as their green image, freight carriers have been trying to come up with innovative solutions such as micro depots and cargo bikes (electric as well as man-powered) (Leonardi et al., 2012).

Autonomous vehicles offer a promising solution to provide efficient delivery services to city dwellers. Technical advances in autonomous delivery vehicles (ADVs) have set the stage for a new era of urban freight delivery. Autonomous trucks or autonomous robots and drones have been tested for last-mile delivery of goods in some urban areas. These innovative technologies may improve efficiency in terms of cost savings as well as reduce the negative environmental impacts and the risk of infection in the pandemic time.

8.2 TYPES OF UN-MANNED VEHICLES IN FREIGHT

8.2.1 Autonomous delivery robots

Autonomous delivery robots (ADRs) are customised, and purpose built unmanned, autonomous vehicles that deliver goods to customers in the last-mile. These can be broadly categorised as side-walk autonomous delivery

DOI: 10.1201/9781003261513-10

robots (SADRs) and on-road autonomous delivery robots (RADRs). As their name suggests SADRs operate on sidewalks and share space with pedestrians. These are usually smaller in size (capacity) and slower in speed to enhance safety. For example, a SADR was used to deliver pharmaceutical items and food from a local restaurant in Fujisawa sustainable smart town, Japan (Panasonic, 2021). On the other hand, RADRs use the road pavement and share space with other vehicles (such as the RADR developed by the tech-company NURO [TechCrunch+, 2020] and the Xiaomanlv used by Alibaba [Alibaba Cloud, 2020]). These have a larger capacity which can be compartmented to carry the demand and serve more than one customer along their routes. For example, Xiaomanlv has a capacity of 50 packages, and it can travel over 100 km on a single charge delivering 500 packages per day (RTIH, 2021).

There are numerous research papers focusing on proving the advantages of the use of ADRs in last-mile delivery based on hypothetical setups and mathematical models, which will be discussed in detail later. For example, Simoni et al. (2020) found that a combined truck and robot delivery system can be more efficient than the truck only system if robots are employed in heavily congested areas. Figliozzi and Jennings (2020) also showed similar results on cost and travel distance between the two systems and also concluded that ADR assisted systems have the potential to significantly reduce energy consumption and CO_2 emissions. Chen et al. (2021) highlighted that contactless delivery to customers in case of ADRs-based delivery systems can minimise the risk of infection during the COVID-19 pandemic.

Apart from the modelled cases, there are many real-life applications of the ADRs in last-mile delivery systems, although most of them are small scale isolated projects or even experimental studies. As mentioned earlier, SADR was used in an experimental operation in Fujisawa sustainable smart town, Japan. These were remotely operated on a pre-determined route (Panasonic, 2021) to validate their technical ability for full-scale systems in future.

Alibaba, the Chinese online retailer announced the use of a fleet of 22 RADRs to deliver 30,000 packages to students, staff and teachers of Zhejiang University, Hangzhou, China for their online event "Double 11" in November 2020 (Alibaba Cloud, 2020). By September 2021, their fleet of 200 RADRs have achieved the landmark of over a million deliveries to 200,000 customers in 52 cities in China (RTIH, 2021). The tech-company NURO was granted authorisation in April 2020 by the U.S. National Highway Traffic Safety Administration to test its R2 RADR on public roads in areas of 9 US cities in Santa Clara and San Mateo counties (Crowe, 2021). Nuro has partnerships within many industries in the grocery, logistics, restaurant, and pharmaceutical sectors. For example, in April 2021, with Domino's, Nuro piloted an autonomous pizza delivery in Houston, and in June same year it partnered with FedEx to test a multi-stop, appointment-based autonomous delivery system (Crowe, 2021).

JD.com is another big tech company, which is operating a large fleet of autonomous robots in many cities of China. Stone (2021) reported that the ADRs of JD.com are equivalent to level 4 of self-driving autonomous vehicles and have been operating on real roads with complicated traffic interactions (such as with traffic signals) under various weather conditions (such as rain). To ensure contactless delivery to hospitals and residential compounds during the COVID-19 outbreak in China, JD's ADRs were used in Wuhan in February 2020. The ADRs delivered more than 13,000 packages and covered more than 6,800 kilometres (Stone, 2021).

8.2.2 Un-manned aerial vehicles

Un-manned aerial vehicles (UAVs) or drones are flying robots which can be remotely controlled and/or fly autonomously. Usually, these drones are light with limited battery life and weight carrying capacity and may be restricted by weather conditions. Nonetheless, many researchers have discussed their advantages over traditional freight transport by trucks/vans. A lot of theoretical and modelling studies are available on the integration of drones with conventional truck-based last-mile delivery systems along with numerous pilot projects demonstrating their practical feasibility. For example, Goodchild and Toy (2018) showed that a drone-based delivery model is much more efficient than traditional truck-based delivery in terms of CO_2 emissions and vehicle-miles-travelled (VMT). In some special conditions, such as in remote areas, and disaster-hit/prone areas, drones can be a very efficient mode of delivery. For example, drones have been used for delivering equipment for emergency medical services (Claesson et al., 2017).

In 2019, Japan's online shopping company Rakuten launched a new island-hopping drone service in partnership with the Japanese supermarket chain Seiyu to Sarushima ("monkey island") situated approximately 1.5 km off the coast of Tokyo Bay. Although the island is inhabited, it is a popular destination with about 200,000 visitors each year for picnics and exploring wartime ruins. With no supermarket available on the island, visitors can order up to 5 kg of products from the partnering supermarket delivered to them via Rakuten's drone delivery service at the cost of 500 Japanese yen (approx. US $4.5) (Rakuten, 2019). Rakuten has been part of many other drone-based delivery tests/services in Japan. For example, in 2017 it collaborated with Lawson convenience stores to deliver hot food items to a remote community going through rebuilding process in the disaster-hit areas of Fukushima, Japan (Rakuten, 2017). Wing Aviation, Australia was among the first few companies which obtained a licence for drone deliveries (CASA, Australia, 2022). Their drones delivered items from supermarkets to nearby customers (within a 10 km radius) and their operations have been expanding over many cities in Australia and throughout the world (Wing.com, 2022). CASA, Australia has given them some relaxations in safety regulations due to their good safety record (CASA, Australia, 2022). Both

Rakuten and Wing Aviation use a class of drones with rotary blades, which makes them capable of hovering and vertical landing. Almost all research that has focused on drone integration in urban delivery assumes the rotary type of drones (Macrina et al., 2020).

Fixed-wing-type drones (*e.g.*, used by Zipline, USA [Zipline website, 2022]) have been used to deliver emergency supplies using parachute mechanism from some time now. For example, drones have been used to conduct deliveries of blood and other medical supplies in remote areas of Rwanda and Madagascar (VOA, 2016). Compared to rotary blades type, fixed-wing-type drones are comparatively easy to maintain and have a larger capacity and range, which makes them ideal for emergency response in far-flanged areas. Recently, Zipline was licensed by the Federal Aviation Administration (FAA), in the United States to deliver medical supplies and personal protective equipment (PPE) kits as part of emergency measures due to the COVID-19 pandemic (BBC, 2020).

8.3 ISSUES RELATED WITH ADVs

8.3.1 Capacity and range

Both ADRs and UAVs face issues of limited capacity and delivery range. For example, the delivery robots of the Starship Company can carry 10 kg of pay load (Hoffmann and Prause, 2018). The high initial cost of ADRs is another issue until mass production is achieved. For example, the research and development costs were over 88 thousand USD/per robot for JD.com, which has steadily reduced to about 7000 USD (Internet of Business, 2022). They further estimate that the operation cost per delivery will become one fifth with ADRs in future.

8.3.2 Regulations

Since freight services using ADRs and UAVs are still in their infancy, regulatory frameworks regarding their operations are also not well defined. For example, in China, autonomous robot delivery services are basically permitted to operate in intelligent/connected demonstration areas such as Shunyi and Yizhuang districts in Beijing and Jiading district in Shanghai. The rules are basically proposed by each district with the cooperation of service operator companies. Recently though, the Beijing High-level Autonomous Driving Demonstration Zone issued their first implementation rules for the operation and management of ADRs in Beijing, China, thereby paving the way to get an official licence to operate ADRs on streets. The rules include standard specifications for the size, load, speed, power and other technical and testing indicators of ADRs. For example, the maximum speed for ADRs has to be less than 15 kmph and they have to intimate other traffic by using light and sound signals at intersections with limited visibility (China News website, 2021). Furthermore, overtaking and driving in reverse are also prohibited. In Japan, the regulations regarding the ADRs have been promulgated by the National Police Agency in June, 2021. The regulations are different for various levels of technology, such as fully automated and semi-remote. The rules include requirements of the experiment objective, restricted area, safety measures, and structure of robot and operator of the robot. For example, robot's length and width should be less than 120 and 70 cm, respectively, and the maximum speed is restricted to only 6 kmph (National Police Agency, Japan, 2021). Furthermore, there is a rule related to information security for the use of information collected by the robots' sensors and camera. The European Union (EU) also has enacted one of the strictest data protection policies that is, the General Data Protection Regulation (GDPR) to return control to citizens and residents over their personal data required to operate (or improve) a delivery robot (including accident avoidance and mitigation) and to run the delivery operations (Hoffmann and Prause, 2018).

In the USA, at least 41 states have considered legislation related to autonomous vehicles by 2012. By 2018, 15 states had enacted 18 AV related bills, depending on the size, weight and speed limit, varying in severity from state to state in the key areas of traffic regulations and product liability in case of an accident. In many states, such as Pennsylvania, Virginia, Idaho, Florida, Wisconsin and Washington, D.C. delivery robots can share the streets with people (Generalcode: website, 2022). San Francisco has one of the most restrictive regulations on SADRs requiring not only speed and weight limits but also require a human chaperone for each device. The maximum number of robots for a company has been set as three with a limit of nine autonomous delivery devices permitted in the city at a time on sidewalks with more than 6 feet width (Wong, 2017). Similar to San Francisco, other high population areas such as New York, are reluctant to adopt ADRs as legal delivery vehicles (Generalcode: website, 2022).

Similarly, UAVs are also subjected to a variety of regulations including a blanket ban of no fly zones in major cities, around airports (mostly 9.3 km radius) and other sensitive facilities such as nuclear power plants (about 19 km). There are also regulations regarding the maximum weight of UAVs, and their minimum altitude (for example 150 m in Japan and Korea). Currently, UAVs are only allowed in day time such as in Australia (CASA, Australia, 2022) and they must remain within visible range for most cases (except the fixed wing type, long-range delivery UAVs mentioned in the previous section).

8.3.3 Acceptability

ADVs are currently subjected to many restrictions and regulations from public authorities. Their acceptance as a new mode of delivery by the customers is also an important issue. Willingness to Pay (WTP) for these new services can be used for the evaluation of public acceptance (Pani et al., 2020). Their survey study in Portland during COVID-19 found that 61.28% responded positively to deliveries using ADRs. They further suggest that competitive pricing and sensitising the public about the advantages of this new technology (*e.g.*, fast delivery speed, contactless handling, and convenience) can motivate customers who were either satisfied with current delivery modes or cited extra costs for their negative responses. Security measures such as theft prevention systems and anti-hacking systems (Humphreys, 2012) can increase the acceptability of un-manned freight vehicles.

Using the Technology Acceptance Model (Davis et al., 1989), which is based on the perceived usefulness and ease of use of technological innovation, Bogatzki and Hinzmann (2020) developed the Autonomous Delivery Vehicle Acceptance Model (ADVAM). Based on a questionnaire survey conducted in Germany via social network services, the ADVAM suggested that if the use of ADVs is convenient, cost effective, exciting, and gives a sense of personal control to customers they will be more likely to accept them. However, they couldn't find a significant connection of feasibility (such as access to internet and cell phones), privacy and security in the acceptance of ADVs, which is interesting as privacy, safety and security are common basis of many regulations relating to ADVs. In a survey conducted among a small group of students at Kyoto University, respondents clearly stated their concerns about privacy and security issues with deliveries using ADRs (Yihe et al., 2020). A majority of students were comfortable with sharing space with SADRs, whereas, about 41% expressed their concern regarding collisions of vehicles with RADRs and the congestion they might cause.

Market surveys suggest that the future use of ADVs will keep increasing as it emerged as the most preferred mode for the surveyed people aged 18–34 years (Joerss et al., 2016). Similar results were obtained in a survey conducted with a small group of students at Kyoto University, where 60% of respondents expressed a positive attitude towards the future use of UAVs as a delivery option (Jee et al., 2020). However, more than 70% showed their concerns about the safety of delivered products.

In an interview, a robot manufacturing company commented that in their own survey about 60% of people didn't mind the movement of delivery robots, 30% paid some attention, whereas 10% had negative views about them (Taniguchi and Qureshi, 2020). They mentioned that safety, external design and communication are some of the areas of concern for people. People don't prefer humanoids, but it is important that delivery robots can communicate with speech and user interface.

Taniguchi and Qureshi (2021) performed an economic analysis of the acceptability of ADRs by considering the experiences of two freight carriers operating in Osaka, Japan. The locations of these customers were extracted from a probe-data set obtained while tracing the route of delivery vehicles using a GPS device. After extracting the actual delivery time (*i.e.*, when a truck stop was identified at their locations) the customers were assigned to different time windows (such as 8 am – 2 pm, 2 pm – 4 pm) based on the actual runs of a traced vehicle. It was assumed that the demand of every customer can be carried by a robot and that robots only deliver to a single customer at a time. With the actual operation of the delivery trucks represented by case 1 ("original"), three other delivery cases were tested based on utilisation of SADRs (such as SADRs only, SADRs and taxi, and SADRs with the mothership and mobile hubs). Furthermore, in order to evaluate the possibilities of full-scale commercial production and adaptation of delivery robots in the delivery industry, various initial costs and operation cost scenarios were also considered. The travel times of robots were estimated as people walking on sidewalks. Using these travel times a bin packing problem was solved considering the length of a time window as the bin capacity to estimate the number of the robots required under each scenario as well as to determine the assignment of the customers to different robots. In case 3, it was assumed that customers with more than 40 minutes of robot travel time will be assigned to delivery via a local taxi service with the objective to raise the earning of taxi companies due to reduced passenger demand during the COVID-19 pandemic. It was found that under a hypothetical future scenario where fixed and operating costs of the ADRs are expected to be reduced due to mass/commercial production of ADRs, a delivery system with mothership and ADRs operated from mobile hubs near the customers' locations would result in similar total delivery cost and services levels (no delays) compared to the actual delivery operations (*i.e.*, case 1: real truck-based delivery observed by probe data).

8.4 INTEGRATION OF ROBOTS AND DRONES IN URBAN FREIGHT DELIVERY SYSTEMS

Due to limited load carrying capacity and range, most of the integrated freight delivery systems consider drones as an assisting partner to the main delivery vehicles. For example, Murray and Chu (2015) introduced the flying sidekick travelling salesman problem (FSTSP) (as shown in Figure 8.1a). In typical FSTSPs one or more drones ride the delivery vehicle along its delivery route; their launching location, customers to be served and retrieving locations are additional optimisation decisions considered along with the vehicle's route. Once launched drones typically deliver to a single customer location and return to the launching vehicle at another stop along its route, thus time synchronisation is a key factor in the FSTSP. In some cases, the delivery vehicle acts as a mothership which can launch multiple drones at a time or relaunch recharged returning drones from a fixed location (Figure 8.1b). Once all the launched drones are returned, the mothership moves to another launching location (*e.g.*, see Moshref-Javadi et al., 2020). In some mothership type models, the stops (or the mobile hub locations) are fixed (such as in Boysen et al., 2018) others include the routing of mothership as an optimisation decision (*e.g.*, Poikonen and Golden, 2019). Usually, minimisation of the total operation time or makespan (service complete time) is considered as the objective function while the drone's limited endurance (such as flying time or battery capacity) is considered as an additional constraint along with typical TSP constraints. Many researchers used distance or time-based endurance for the drones, however, Raj and Murray (2020) showed that drones can be better utilised if their speed is controlled while considering their endurance based on speed and parcel load.

On the other end of the integration spectrum, the drone delivery problem (DDP) (Dorling et al., 2017) only considers drones launched from a fixed depot as delivery agents (as shown in Figure 8.1c). Limited range drones may require consideration of charging stations in the DDP (such as in Coelho et al., 2017) or else the solution may become infeasible. The parallel drone scheduling travelling salesman problem (PDSTSP) (Murray and Chu, 2015) (Figure 8.1d) solves this issue by combining a TSP (for a conventional delivery vehicle) with the DDP. Although the assignment of customers considers both drones and the delivery vehicle, simultaneously, based on their range, their routing (and scheduling) is independent of each other and therefore no time synchronisation is required between the two operations. For example, Lee et al. (2022) presented a realistic case study of deliveries from a supermarket in South Korea based on the PDSTSP.

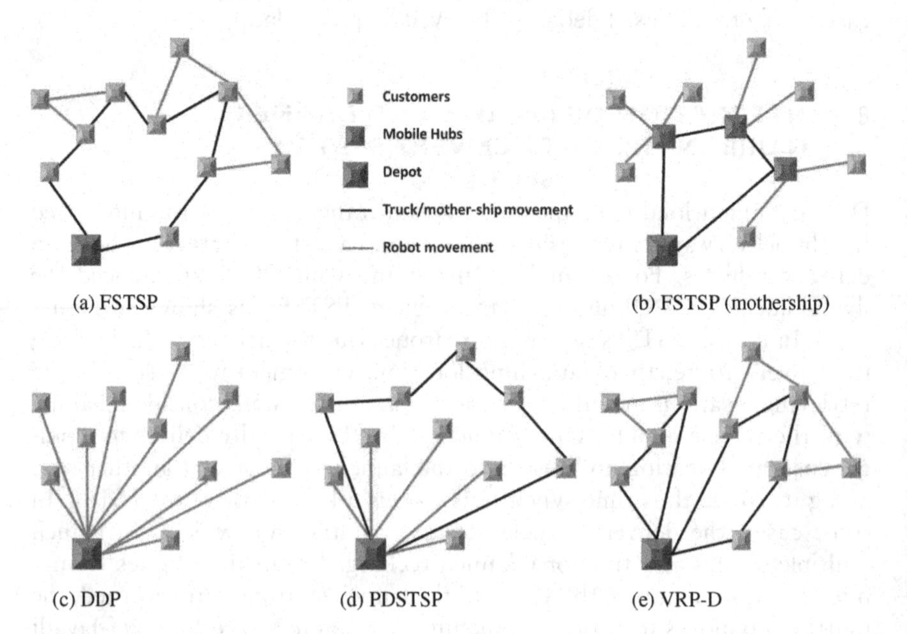

Figure 8.1 Integration of the drones in the vehicle routing.

The vehicle routing problem with drones (VRP-D) adds the capacity constraint for vehicles and therefore may have multiple truck routes (Figure 8.1e). Many researchers have considered an FSTSP type of route for each delivery vehicle in the VRP-D (e.g., Sacramento et al., 2019), however, Schermer et al. (2019) introduced the idea of cyclic deliveries by drones where the delivery vehicles act similar to a mothership. Di Puglia et al. (2020) considered the time windows version of the VRP-D, whereas, Chiang et al. (2019) introduced a green VRP-D by including minimisation of the CO_2 emissions as an objective function. Liu et al. (2020) used an energy consumption model that depends on the payload as well as on the distance travelled to determine a drones' range in their two-echelon VRP-D. ADRs are being integrated into the classic delivery models in more or less the same way as drones.

8.5 INDOOR LOGISTICS

For outside deliveries, the use of ADRs is facing challenges such as traffic regulations, safety concerns (especially with pedestrians), security of goods and robots themselves, and challenges of difficult terrain and weather conditions. Their use in a closed indoor environment presents many feasible conditions, and that's why there are more use cases in such environments (Taniguchi and Qureshi, 2020). Robots with "follow me" mode are being deployed in workplaces to carry instruments and equipment for the technicians; in future, repair/maintenance workers can ride public transport with such robots to offer door-to-door services (Taniguchi and Qureshi, 2020). Other indoor uses of robots can be found in restaurants and hospitals; for example, Kyoto University Hospital uses HOSPI to transport samples from the blood collection room to the laboratory (Panasonic [https://connect.panasonic.com/jp-ja/case-studies/kyoto-university-hospital]). By controlling the two units in a group, the transportation efficiency is improved in a well-balanced manner. It can be operated without a driving guide, but in consideration of patient safety, a sticker is attached to call attention to the passage of HOSPI. HOSPI takes evasive action when it is about to collide with a person or an object.

8.6 CONCLUSIONS

Introduction of advanced delivery technologies such as autonomous delivery robots and drones in city logistics has been a topic of research interest for quite some time in academia. Various models integrating these technologies have been presented but still, the number of practical applications is limited, especially at a level where substantial share is delivered by drones or autonomous delivery robots. Their current technology and design still weigh down them from the view points of load carrying capacity and range. Furthermore, administrators are still reluctant to relax the current regulations related to their operations. Advancements in technology and increasing evidence from pilot and small scale practical applications are paving the way for their wider acceptance among stakeholders such as freight carriers and administrators. The COVID-19 pandemic has also provided an encouraging environment for the accelerated integration of these technologies in city logistics practices. Therefore, it is expected to observe a continued increase in the trend of use of these technologies both in research and practice.

REFERENCES

Alibaba Cloud (2020). Alibaba launches robot-only delivery service for double 11. Available at https://www.alibabacloud.com/blog/alibaba-launches-robot-only-delivery-service-for-double-11_596954 (accessed on March 3, 2022).

BBC (2020). Zipline drones deliver supplies and PPE to US hospitals. Available at https://www.bbc.com/news/technology-52819648 (accessed on March 4, 2022).

Bogatzki, K. and Hinzmann, J. (2020). *Acceptance of Autonomous Delivery Vehicles for Last Mile Delivery in Germany. Extension of the Technology Acceptance Model to an Autonomous Delivery Vehicles Acceptance Model*, Masters thesis, Jonkoping International Business School, Jonkoping University, Sweden.

Boysen, N., Briskorn, D., Fedtke, S. and Schwerdfeger, S. (2018). Drone delivery from trucks: Drone scheduling for given truck routes, *Networks*, **72**, 506–527.

CASA, Australia (2022). Drone delivery services, Civil Aviation Safety Authority (CASA), Australian Government. Available at https://www.casa.gov.au/drones/industry-initiatives/drone-delivery-services (accessed on March 4, 2022).

Chen, C., Demir, E., Huang, Y. and Qiu, R. (2021). The adoption of self-driving delivery robots in last mile logistics, *Transportation Research Part E*, **146**, 102214.

Chiang, W.C., Li, Y., Shang, J. and Urban, T.L. (2019). Impact of drone delivery on sustainability and cost: Realizing the UAV potential through vehicle routing optimization, *Applied Energy*, 242, 1164–1175.

China News Website (2021). Available at https://www.chinanews.com.cn/cj/2021/05-25/9485293.shtml (accessed on March 24, 2022).

Claesson, A., Bäckman, A., Ringh, M., Svensson, L., Nordberg, P., Djärv, T. and Hollenberg, J. (2017). Time to delivery of an automated external defibrillator using a drone for simulated out-of-hospital cardiac arrests vs emergency medical services. *JAMA*, 317, 2332–2334.

Coelho, B.N., Coelho, V.N., Coelho, I.M., Ochi, L.S., Haghnazar, K.R., Zuidema, D., Lima, M.S.F. and da Costa, A.R. (2017). A multi-objective green UAV routing problem, *Computers and Operations Research*, **88**, 306–315.

Crowe, S. (2021). Nuro raises $600M for autonomous delivery vehicles. Available at https://www.therobotreport.com/nuro-raises-600m-autonomous-delivery-vehicles/ (accessed on March 3, 2022).

Davis, F.D., Bagozzi, R.P. and Warshaw, P.R. (1989). User acceptance of computer technology: A comparison of two theoretical models, *Management Science*, **35**, 982–1003.

Di Puglia Pugliese, L., Guerriero, F. and Macrina, G. (2020). Using drones for parcels delivery process, *Procedia Manufacturing*, **42**, 488–497.

Dorling, K., Heinrichs, J., Messier, G.G. and Magierwski, S. (2017). Vehicle routing problem for drone delivery, *IEEE Transactions on Systems, Man, and Cybernetics: Systems*, **47**, 1–16.

Figliozzi, M. and Jennings, D. (2020). Autonomous delivery robots and their potential impacts on urban freight energy consumption and emissions, *Transportation Research Procedia*, **46**, 21–28.

Generalcode: Website (2022). Autonomous robot delivery legislation. Available at https://www.generalcode.com/blog/autonomous-robot-delivery-legislation/ (accessed on June 11, 2022).

Goodchild, A. and Toy, J. (2018). Delivery by drone: An evaluation of unmanned aerial vehicle technology in reducing CO_2 emissions in the delivery service industry, *Transportation Research Part D*, **61**, 58–67.

Hoffmann, T. and Prause, G. (2018). On the regulatory frame-work for last-mile delivery robots, *Machines*, **6**, 6–8.

Humphreys, T. (2012). *Statement on the Vulnerability of Civil Unmanned Aerial Vehicles and Other Systems to Civil GPS Spoofing*, University of Texas at Austin, 1–16.

Internet of Business (2022). JD.com launches robot delivery in China. Available at https://internetofbusiness.com/jd-com-robot-delivery-china/ (accessed on March 24, 2022).

Jee, H., Lai, Y. and Lin, X. (2020). *Advances in Freight Transport: Challenges and Opportunities, Capstone Project Report*, Department of Urban Management, Kyoto University, Kyoto, Japan.

Joerss, M., Schroder, J., Neuhaus, F., Klink, C. and Mann, F. (2016). *Parcel Delivery: The Future of Last Mile*, McKinsey & Company. Available (in Japanese) at https://bdkep.de/files/bdkep-dateien/pdf/2016_the_future_of_last_mile.pdf (accessed on March 3, 2022).

Lee, S.Y., Han, S.R. and Song, B.D. (2022). Simultaneous cooperation of refrigerated ground vehicle (RGV) and unmanned aerial vehicle (UAV) for rapid delivery with perishable food, *Applied Mathematical Modelling*, **106**, 844–866.

Leonardi, J., Browne, M. and Allen, J. (2012). Before-after assessment of a logistics trial with clean urban freight vehicles: A case study in London, *Procedia - Social and Behavioral Sciences*, **39**, 146–157.

Liu, Y., Shi, J., Wub, G., Liu, Z. and Pedrycz, W. (2020). Two-echelon routing problem for parcel delivery by cooperated truck, *IEEE Transactions on Systems, Man, and Cybernetics: Systems*, **99**, 1–16.

Macrina, G., Pugliese, L.D.P., Guerriero, F. and Laporte, G. (2020). Drone-aided routing: A literature review, *Transportation Research Part C*, **120**, 102762.

Moshref-Javadi, M., Hemmati, A. and Winkenbach, M. (2020). A truck and drones model for last-mile delivery: A mathematical model and heuristic approach, *Applied Mathematical Modelling*, **80**, 290–318.

Murray, C.C. and Chu, A. (2015). The flying sidekick traveling salesman problem: Optimization of drone-assisted parcel delivery, *Transportation Research Part C*, **54**, 86–109.

National Police Agency, Japan (2021). Road use permission standards related to public road demonstration experiments such as specified automatic delivery robots. Available (in Japanese) at https://www.npa.go.jp/bureau/traffic/selfdriving/robotkijun2.pdf (accessed on March 24, 2022).

Panasonic (2021). Press release (in Japanese). Available at https://news.panasonic.com/jp/press/data/2021/03/jn210304-1/jn210304-1.html (accessed on March 3, 2022).

Pani, A., Mishra, S., Golias, M. and Figliozzi, M. (2020). Evaluating public acceptance of autonomous delivery robots during COVID-19 pandemic, *Transportation Research Part D*, **89**, 102600.

Poikonen, S. and Golden, B.L. (2019). The mothership and drone routing problem, *INFORMS Journal on Computing*, **32**, 199–530.

Raj, R. and Murray, C. (2020). The multiple flying sidekicks traveling salesman problem with variable drone speeds, *Transportation Research Part C*, **120**, 102813.

Rakuten (2017). Rakuten Drone makes convenience store deliveries in Fukushima. Available at https://rakuten.today/blog/rakuten-drone-delivers-hot-meals-fukushima.html (accessed on March, 4, 2022).

Rakuten (2019). Rakuten's new island-hopping drone delivery service. Available at https://rakuten.today/tech-innovation/island-hopping-drone-delivery-service.html (accessed on March 4, 2022).

RTIH (2021). Alibaba Group Xiaomanlv robots hit e-commerce delivery milestone. Available at https://retailtechinnovationhub.com/home/2021/9/29/alibaba-group-xiaomanlv-robots-hit-e-commerce-delivery-milestone (accessed on March 3, 2022).

Sacramento, D., Pisinger, D. and Ropke, S. (2019). An adaptive large neighborhood search metaheuristic for the vehicle routing problem with drones, *Transportation Research Part C*, **102**, 289–315.

Schermer, D., Moeini, M. and Wendt, O. (2019). A matheuristic for the vehicle routing problem with drones and its variants, *Transportation Research Part C*. **106**, 166–204.

Simoni, M.D., Erhan Kutanoglu, E. and Claudel, C.G. (2020). Optimization and analysis of a robot-assisted last mile delivery system, *Transportation Research Part E*, **142**, 102049.

Stone, T. (2021). Case study: China's JD.com launches fleet of 30 new autonomous delivery bots. Available at https://www.traffictechnologytoday.com/news/autonomous-vehicles/case-study-chinas-jd-com-launches-fleet-of-30-new-autonomous-delivery-bots.html (accessed on March 4, 2022).

Taniguchi, E. and Qureshi, A.G. (2020). Interview with a robot manufacturing company, conducted on September 17, 2020.

Taniguchi, E. and Qureshi, A.G. (2021). Emergency urban delivery management during the COVID-19 pandemic, report submitted to the Kinki Kensetsu Kyokai (in Japanese).

VOA (2016). Voice of America article, Drones Helping to Save Lives in Rwanda and Madagascar. Available at https://learningenglish.voanews.com/a/how-drones-are-helping-save-lives-in-rwanda-and-madagascar/3559871.html (accessed on March 24, 2022).

Wing.com (2022). Available at https://wing.com/, (accessed on March 4, 2022).
Wong, J. (2017). San Francisco sours on rampant delivery robots: Not every innovation is great. Available at https://www.theguardian.com/us-news/2017/dec/10/san-francisco-delivery-robots-laws (accessed on March 24, 2022).
Yihe, Z., Yufeng, G., Namulindwa, S., Jinchuan, Z., Xihan, L. and Yingzhou, S. (2020). Assessing the Social Acceptability of Sidewalk Autonomous Delivery Robots During COVID-19, Capstone Project Report, Department of Urban Management, Kyoto University, Kyoto, Japan.
Zipline Website (2022). Available at https://flyzipline.com/technology/ (accessed on May 4, 2022).

Chapter 9

Access management and pricing

9.1 INTRODUCTION

This chapter presents a range of analytics methods that illustrate how big data, analysis tools, and models can improve pricing and access management schemes for freight vehicles in urban areas. Descriptive analytics procedures presented include monitoring usage of loading bays and loading docks. Prescriptive analytics methods described include models developed for optimising the number of bays in loading docks and on-street as well as toll levels for trucks.

9.2 UNLOADING BAYS

Unloading bays or loading zones are designated on-street areas that allow freight vehicles to park to unload or load goods that enhance efficiency and safety. When planning and managing loading zones governments need to make decisions related to their location, capacity, duration limits, type of availability information and whether they should allow them to be reserved. It is important to evaluate the effects of decisions based on surveys or models. A number of criteria can be used to investigate the impacts of loading zone provision and management. Performance measures relating to the operations of freight vehicles such as the distance travelled by vehicles as well as walking times, vehicle utilisation and productivity are important to consider. Loading zone schemes can also affect overall congestion levels as well as emissions and fuel consumption.

9.2.1 Monitoring usage

There are a range of sensor technologies, including RFID, GPS and CCTV that can be used for data acquisition for supporting unloading bay telematics systems (Kijewska, 2018). Surveillance technologies can provide dynamic real-time information relating to the occupancy of unloading bays

DOI: 10.1201/9781003261513-11

(Chloupek, 2012). Video and access information systems can be used to analyse the arrival and duration characteristics of freight vehicles at unloading bays (Chiara and Cheah, 2017).

A number of interactive mapping tools have been developed by Transport for New South Wales (TfNSW) relating to on-street loading zones in Sydney's CBD. Maps can be produced displaying loading zones in proximity to the delivery destination for a specified service duration as well as time period and day of week (Interactive Maps). Another mapping tool allows details of the demand of loading zones to be displayed for specified zones, streets, areas and time periods. Historical "push button" data from ticket machines provide a large database that allows for the temporal and spatial analysis of demand throughout the CBD to be undertaken.

Figure 9.1 shows the historical daily demand profile for a specific loading zone (Sydney CBD Loading Zone Usage). This tool can be used by carriers to plan more efficient delivery routes. Administrators can also use this system for planning and managing loading zones. For example, analysis of temporal demand patterns has revealed significantly higher demand in morning periods for many loading zones that has led to the government promoting more deliveries in the afternoon (Stokoe, 2019).

A comparison of the performance of various transport modes was conducted to investigate courier trip productivity in Sydney's CBD (Stokoe, 2017). Delivery exercises were undertaken for assessing routes conducted

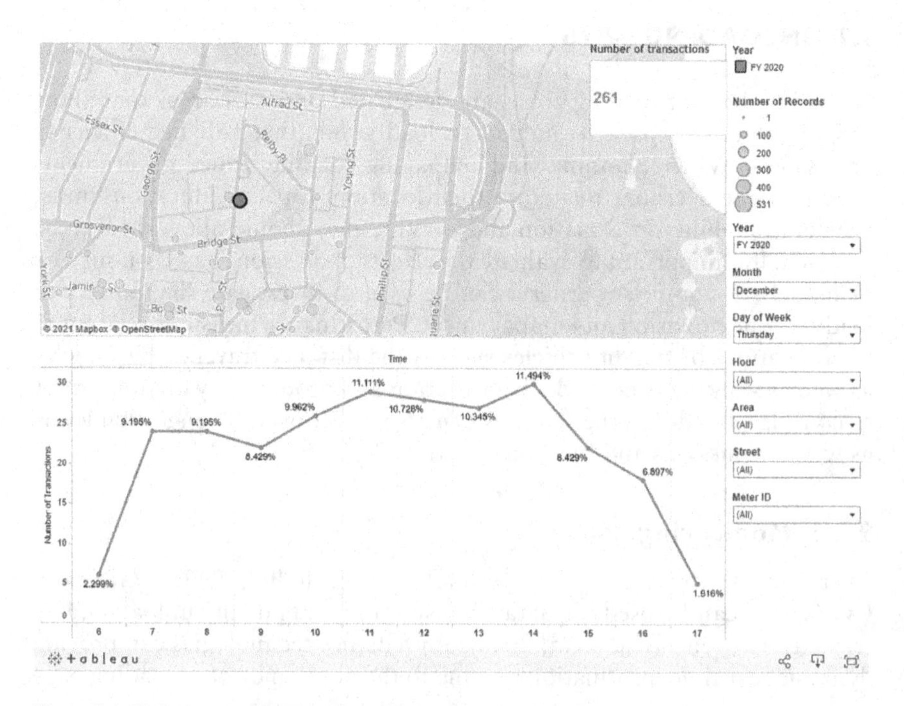

Figure 9.1 Utilisation of loading bays (Transport for NSW).

by vans, walkers and bikes. GPS devices were used to measure times and distances travelled by van, bike and walking, times taken as well as loading zone dwell times.

9.2.2 Models

Iwan et al. (2018) presented a model for predicting the reduction of air pollution by introducing unloading bays in Szczecin and Oslo. Since speeds less than 50 km/h typically have much higher emissions levels per kilometre, loading bays were shown to significantly reduce emissions by increasing traffic speeds.

A model for determining the optimal location of loading bays considering delay penalties, operating costs, parking fees and delays for delivery vehicles and passenger vehicles was presented by Aiura and Taniguchi (2005). The model was able to reduce total costs of the configuration of loading bays by 16%. An investigation of benefits based on the violation levels of passenger cars using unloading bays was conducted.

A microsimulation model that allows the impacts of double-parking of freight vehicles to be predicted was used to examine how enforcement and the optimal location of loading bays can improve traffic flow in Lisbon (Alho et al., 2018). This model predicted that having optimal bay locations and enforcement of usage leads to improved traffic flow and subsequently better mobility on the road network.

A model for optimising the number and location of loading bays was presented by Letnik et al. (2018). When applied to Lucca, significant savings in distance and time travelled for freight vehicles as well as reductions in fuel consumption and CO_2 emissions were predicted with the optimised configuration loading bay configuration.

A multi-objective approach was developed to optimise courier routes in CBDs that consider both driving and walking routes (Thompson & Zhang, 2018). An optimisation model was developed for determining the best routes for minimising operating and environmental costs. A driving coefficient was used to express the relative weighting of driving distances between loading zones compared with the weighting of walking distances for walking routes from unloading bays (loading zones). Optimal routes were determined for delivering parcels in Melbourne's CBD. The influence of the driving coefficient on driving and walking distances was analysed to investigate the effects on environmental costs for various duration limits at loading zones. Increasing the driving coefficient means enhancing the priority of reducing driving in the delivery route optimisation, leading to reduced driving distances and times. It was shown that reducing parking duration limits in loading zones leads to higher environmental costs. This model was extended and applied to study the effects of introducing loading zone occupancy information as well as estimate its potential limitations compared to a reservation system (Zhang & Thompson, 2019).

Advances in on-line booking systems and sensor technologies can provide real-time information relating to the availability and occupancy levels at loading zones. These have the potential to improve the efficiency of courier routes in central city areas. An agent-based simulation model combined with optimisation procedures has been developed to predict the behaviour of courier drivers in CBDs to study the benefits of introducing loading zone occupancy information as well as estimate the potential limitations compared to a reservation system (Zhang & Thompson, 2019). Courier routes were optimised based on minimising both driving and walking costs. The consequences of permitting waiting at loading zones as well as duration limits were predicted.

When no waiting was allowed, lower driving times were estimated when real-time availability and average occupancy information of loading zones were provided. This was found to be more pronounced when there were no duration limits at loading zones. On the other hand, when waiting at loading zones was permitted, only marginal driving time savings were predicted when real-time availability and average occupancy information were provided. Compared to the case when an ideal reservation system was employed for couriers to conduct the best possible delivery plan, the availability information systems still had opportunities for further enhancements in driving times during periods of high demand, although only marginal savings in driving times were predicted when waiting at loading zones were allowed.

Optimisation procedures have also been developed to investigate the relationship between reliability and financial performance for retail swap networks, with unreliability measured by the number of transfers made (Zhang & Thompson, 2021). Figure 9.2 shows the trade-off between vehicle distances travelled (VKT) and Load Factor (LF) and the maximum

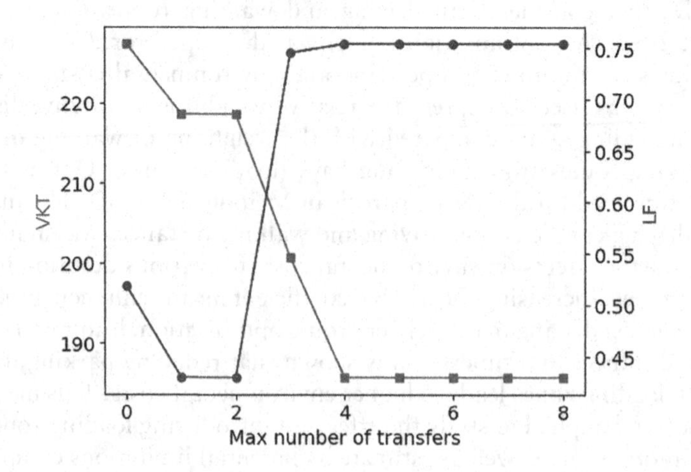

Figure 9.2 Trade-off analysis.

number of transfers permitted in a simple retail network where goods need to be swapped between stores.

A model to predict the benefits of an unloading bay booking system in Winchester, UK highlighted the need for a range of performance measures to be considered including the adherence to booking, delivery time and the possibility of this system to spread peak demand and overall congestion (McLeod & Cherrett, 2011).

9.3 LOADING DOCKS

There is a need for descriptive and predictive analytics for understanding and forecasting freight demand at major activity hubs in urban areas. If an insufficient number of off-street loading are provided and roadside space is not available, buildings become challenging to serve, inefficient for deliveries, and less attractive to customers, resulting in impacts on the urban environment.

The provision of loading dock capacity during the design and planning processes and development approval is often contentious. In central city areas where land values are high, servicing capacity is often a cause of disagreement between planning authorities and developers. There is also a low level of confidence in current guidelines used in planning processes in many cities due to their reliance on outdated data and inadequate articulation of all the considerations needed to provide suitable facilities which support existing processes.

9.3.1 Surveys

Surveys can be undertaken at loading docks to understand the demand patterns at major generators. An observational based survey was conducted to record details of the deliveries and pick-ups at a tower in Melbourne's CBD (Thompson & Flores, 2016). The arrival and departure times of vehicles, type of vehicle (car, van or truck), name of carrier, type of goods and the level of tenant were captured. This allowed arrival profiles and the frequency distribution of dwell times for vehicle types to be constructed.

An investigation of the correlation between arrivals, duration and vehicles types was also undertaken. The relationship between dwell time or duration and vehicle type was shown to be not independent with trucks being overrepresented in the high duration category and cars being overrepresented in the low duration category. The association between arrival period and type of goods category was found to be not independent with food goods being overrepresented in arrivals in the morning business hours period, fruit being overrepresented in arrivals during the early period and couriers in the late period. However, no significant relationship was found between the arrival period and vehicle type as well as the arrival period and duration category. There was also no significant relationship identified between goods type and vehicle type as well as goods type and duration category.

9.3.2 Capacity modelling

A Decision Support System (DSS) for forecasting the freight generated for developments and determining the optimal provision of on-site loading docks for new major developments in Sydney has recently been developed (Aljohani et al., 2021).

The model processes various parking surveys collected by Transport for NSW (TfNSW) from buildings in Sydney for different land uses. Regression and clustering analysis techniques were used to develop the predictive modelling methodology. The model development approach ensures a mathematically robust process to ensure the outputs' validity based on the observed datasets.

Two interactive templates form the output. An 'Optimisation Solver' template determines the recommended dock configuration for the building under consideration by calculating the optimal number of unloading bays. The model minimises the parking area while keeping the dock's effectiveness (ability to accommodate incoming vehicle demand) to a user configurable service level. A second template, the 'Dashboard' presents details of the predicted parking demand, vehicle movements and utilisation of the dock. The Dashboard is an interactive and transferrable template that various stakeholders could use in different locations to input the parameters and generate results and outputs.

Microsoft Excel's Solver Add-in has been used to generate a solution as illustrated for a building shown in Figure 9.3. An algorithm considers various combinations of small, medium and large spaces to identify an optimal solution meeting the required efficacy level of meeting a certain threshold of parking demand. Simulated demand for a building with user-specified characteristics is carried out and then assessed to the extent to which a given dock configuration can meet the simulated demand. Peak demand is determined based on the regression coefficients and the building characteristics. Peak demand (maximum daily demand) is highly dependent on building characteristics. Best-fitting regression models were calculated to allow for peak demand prediction for small, medium, and large vehicles based on building characteristics. Forecasts are strongly associated with real data and make sense from a domain knowledge perspective.

Overall, the optimal dock configuration recommended by the model not only satisfies the estimated parking demand but also minimises the overall area of the dock and other supporting regulations such as a specified minimum number of loading spaces for medium and/or large vehicles.

The DSS also provides additional insights into a building's freight and servicing activity levels and loading dock operations and efficiency through

FILL IN THE YELLOW CELLS B4:B10 and B14:B24

BUILDING CHARACTERISTICS		
INPUT	VALUE	COMMENT
Number of floors	26	
Commercial area, m2	23000	
Residential area, m2	0	
Number of apartments		leave empty if unknown (for a primarily commercial building)
Retail area, m2	500	
Availability of a dedicated elevator	no	if uncertain, select "no"
Primary use type	commercial	

PEAK DAILY DEMAND	
Vehicle class	Peak demand
Small	10.1
Medium	2.2
Large	1.4

ANALYSIS SETTINGS		
PARAMETER	VALUE	COMMENT
Minimum required efficacy (average % of vehicles to be accomodated during the day)	80	Optimal numbers of small, medium and large spaces are found by minimising the total area of the parking lot while maintaining minimum efficacy (average % of vehicles to be accomodated during the day) at least at this level
Area of 1 small space	18	m^2
Area of 1 medium space	24.5	m^2
Area of 1 large space	31.5	m^2
Minimum number of small spaces	0	increase if you find the optimal number to be unacceptably small
Minimum number of medium spaces	0	
Minimum number of large spaces	0	
Maximum number of small spaces	100	decrease if you find the optimal number to be unacceptably large
Maximum number of medium spaces	100	
Maximum number of small spaces	100	
Maximum total number of spaces	100	

OPTIMAL GENERATED SOLUTION	
SIZE	OPTIMAL NUMBER
SMALL	5
MEDIUM	1
LARGE	0
SOLUTION'S CHARACTERISTICS	
Total spaces	6
Average accomodated vehicles	2.0
Average demand	2.4
EFFICACY (average % of vehicles to be accomodated during the day)	82.4
TOTAL PARKING AREA, m^2	114.5

USER-SPECIFIED SCENARIO	
SIZE	NUMBER
SMALL	5
MEDIUM	0
LARGE	0
SOLUTION'S CHARACTERISTICS	
Total spaces	5
Average accomodated vehicles	1.7
Average demand	2.4
EFFICACY (average % of vehicles to be accomodated during the day)	72.4
TOTAL PARKING AREA, m^2	90

Figure 9.3 Loading dock model outputs.

several parking characteristics (Figure 9.4). This increases understanding of the performance of operations, dock occupancy, peak requirements and parking availability throughout the day. For instance, the efficacy of different dock configurations can be compared with the likely success of the recommended loaded spaces in meeting different parking demand thresholds. Additionally, the highest and lowest hour intervals are identified in terms of highest parking demand, vehicle arrivals by class, activity visits and success and/or failure to accommodate incoming vehicles.

The DSS provides various stakeholders including transport authorities, city planners and property developers, with a tool to assess the requirements in advance during the planning and approval process of new developments. Applications of the model include space proofing, supporting planning applications, enhancing the overall logistics delivery and service operations for new development, and streamlining traffic flow in and around a development, making it more attractive for tenants and end-users.

9.3.3 Loading dock booking systems

Loading dock booking systems can address a number of delivery problems that are common at large activity hubs (Sanders et al., 2016). Unscheduled and unannounced arrival by freight vehicles at loading docks can have many

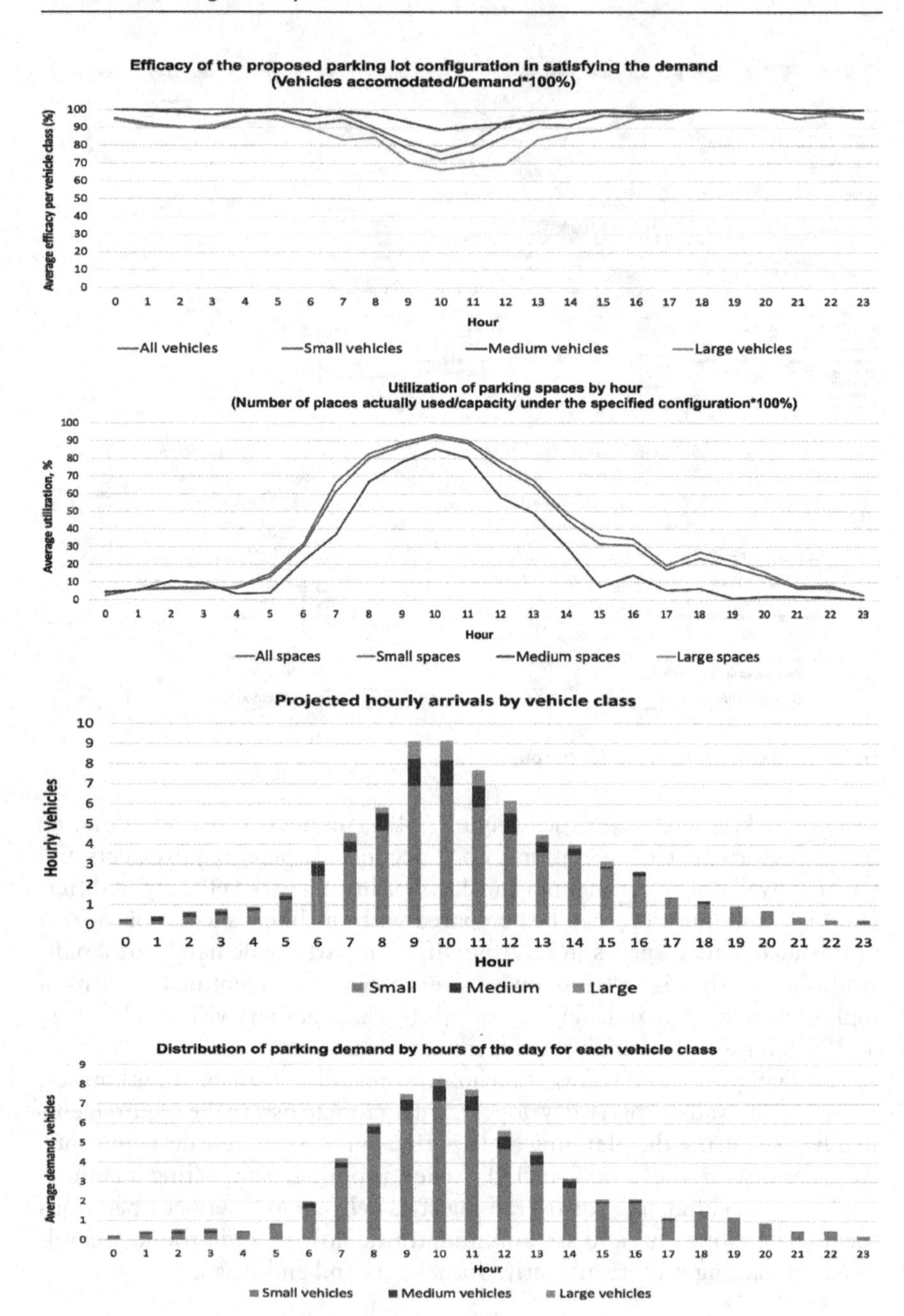

Figure 9.4 Snapshot of the dashboard outputs.

consequences. These relate to the substantial costs that are associated with vehicles queuing at entrances to loading docks due to all loading/unloading bays being occupied during peak periods. This can result in increased traffic congestion on adjacent streets. The time spent coordinating deliveries manually including communication between suppliers, carriers, buyers and receivers can also be substantial.

The MobileDOCK booking system has been implemented at the Emporium in Melbourne's CBD. It was considered important by managers of the Emporium to smooth the delivery patterns throughout the day since they do not like too many drivers in the complex at one time, producing clutter and conflicts with shoppers. Significant safety issues can arise from too many vehicles trying to enter and leave a facility. Security can also be a concern in basement loading docks with some vehicles using the facility without delivering goods to that facility.

Surveys can be undertaken to measure the dwell times of vehicles and queue lengths. The arrival times and durations of freight vehicles by vehicle type, goods carried, shipper and receiver also provide information that is helpful to understand the demand patterns at a facility.

The main goal at the Emporium in Melbourne was to develop a solution for spreading the delivery peaks to improve amenity, safety and security. Objectives included minimising delays and dwell times of carriers. Criteria used considered the time vehicles spend in the queue as well as the number of crashes in the facility.

It is common practice to allow casual arrivals of freight vehicles at loading docks where the bays can be accessed by delivery vehicles on a first come, first served basis in many large loading docks. However, there are generally a number of options available to facility managers to help achieve their goals. Other initiatives that have the potential for improving the operational performance of loading docks include, opening the facility for longer hours to permit off-hour deliveries, mandating higher load factors as well as only permitting deliveries from an Urban Consolidation Centre (UCC) using a Joint Delivery Service (JDS).

Simulation models can be developed to predict queue lengths and delays as well as the number of loading bays required in the facility based on the number and type of stores in the facility. Predictions of revenue from bookings are also required for budgeting capital and operational costs.

Evaluation of the options typically involves consideration of the financial costs, environmental impacts and revenue. Booking systems require software systems to be purchased and operated by facility managers. Revenue is generated by MobileDOCK for each booking made and this is paid by the facility owner. The facility owner typically passes this cost onto the receivers (tenants) as a service fee. Options could be assessed on their environmental performance by considering the impacts on fuel consumption and emissions.

Implementation of a booking system entails the facility management purchasing and operating software systems for creating and managing bookings as well as controlling access to the facility by installing gates and control systems. Payment systems also need to be established.

After implementing the system, reviewing the performance of a booking system can involve monitoring dwell times as well as the arrival times of vehicles compared with the booked times. The number of "no shows" are also of interest for determining how the system is performing.

Bookings can be made over the internet or using a mobile phone. Booking data allows a wide variety of analysis to be undertaken. Details of the appointments include the time, carrier, receiver (*e.g.*, establishment), dock area, door number, service time, number of items (small and large) and status (approved, arrived, departed). Dock managers can link appointment data to databases of suppliers, carriers, buyers and receivers. Utilisation profiles by time of day and day of week can be produced. Deliveries can be linked to the type of receiver to identify temporal patterns of total daily deliveries (Figure 9.5).

Booking details can be linked to vehicle arrival and departure data to analyse supplier and carrier performance as well as the overall centre efficiency. These include the percentage of "no shows", the percentage of on time arrivals as well as the percentage of booked deliveries. Matrices categorising the percentage of deliveries arriving on site (very early, early, on time, late and very late) by the departure time (very early, early, as expected, late and very late) can also be produced. Procedures have been developed to assess the performance of booked deliveries to report timeliness and carrier arrivals, and time on site, highlighting late arrivals and slow load/unload times (Figure 9.6).

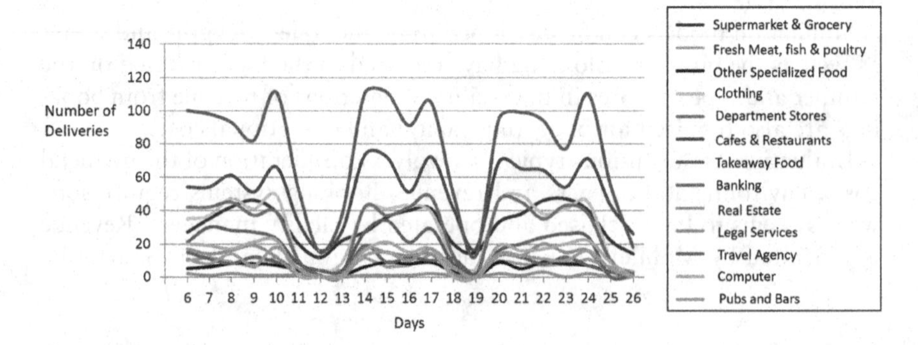

Figure 9.5 Freight generation using loading dock booking data.

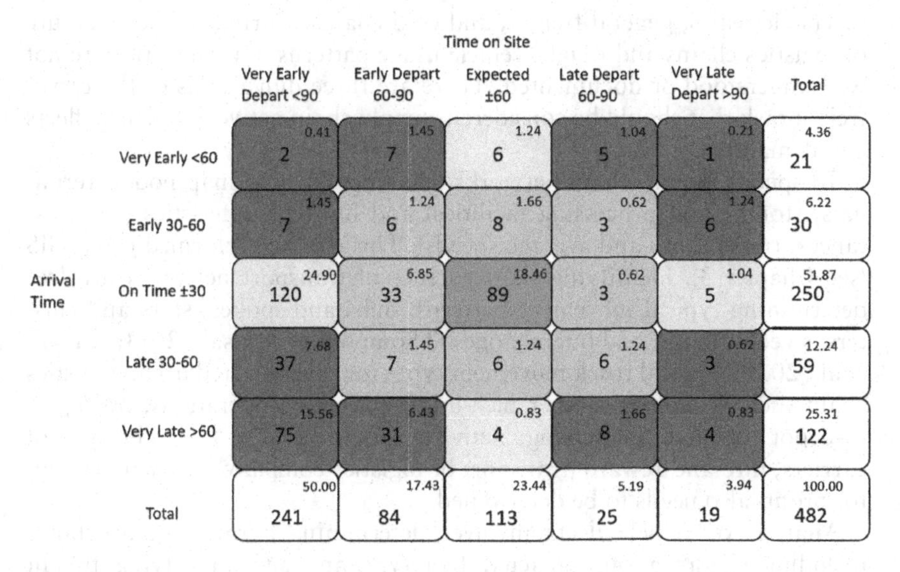

Figure 9.6 Analysis of vehicle arrival and departure times at loading docks.

9.4 ACCESS CONTROL SCHEMES

It is important that access control measures in urban areas are consistent with the cities management objectives (Browne et al., 2005). Often access control schemes lead to additional costs for transport operators. It is important that the benefits and any increased costs are identified and estimated. City access schemes are applied and introduced in different ways according to the objectives an access scheme aims to achieve. Browne et al. (2008) present a classification of access and loading restrictions as well as an overview of how these schemes were initially implemented in European cities. Currently, there are over 700 urban vehicle access regulations operating in 24 European countries (www.lowemissionzones.eu). Quak (2015) provides an overview of common types of access restrictions in urban areas for freight vehicles describing their objectives, effects and enforcement issues.

Pache and Morel (2021) describe how a low-emission zone was established in Lyon. De Marco et al. (2018) found that E-commerce as a proportion of GDP and share as well as GDP per capita were influencing factors affecting the implementation of City Logistics regulations in European cities.

To predict the effect of access control schemes it is necessary to estimate the impact on freight operations for various types of urban supply chains. Urban logistics chains are complex networks that have distinct characteristics that are influenced by the type of goods and the location of terminals, logistics facilities and customers. Common urban logistics chains include shipping containers, construction materiel, bulk liquid fuels, reverse logistics, general freight, and cold chains. Currently, the structure of logistics chains and freight vehicle usage patterns in many cities are not well understood or documented. There are three dimensions of the urban freight task that should be considered, supply chain networks, vehicle fleets and demand.

To specify supply chain networks it is important to map nodes (terminals, storage and processing facilities) and link characteristics (*e.g.*, distances, travel times and average speeds). This can be performed using GIS (see Chapter 3). Identifying the structure of transport networks involves determining typical movement patterns: hubs and spokes, stars and clusters, sweeps, loops, and interchanges (Thompson & Hassall, 2014; Hassall et al., 2019). Typical truck movement types for each selected urban logistics chain such as multi-customer pick-up or put-down operations, or depot-to-depot transfers (interchange activity) should be identified. The type of carriers, hire-and-reward (transport or logistics company) or ancillary (not for profit) also needs to be determined.

Analysis of vehicle fleets involves determining a range of attributes including classification, capacity, fuel type and age. Specifying freight demand involves determining the time windows of receivers and seasonality patterns.

Current vehicle operations can be analysed using vehicle movement data can be captured from telematics systems or GPS (see Chapter 3). Automated data streams from fleet operations from accessing GPS and telematics data from truck cabins and trailers can be analysed. This should be used as a baseline for estimating the impacts on freight vehicles by selected options.

Network criteria including task travel times and vehicle numbers should be estimated. Typical metrics that reflect freight costs are total kilometres travelled, total hours of operation and individual fleet vehicle numbers can be gained from analysis of vehicle telematics reports and data. Typical queuing delays, degree of empty/un-laden returns and directional vehicle trip kilometres can also be estimated.

An analysis of on duty hours, average km per trip duty, driving shift times (terminal, depot or retailer operating hours) and operation hours should be conducted. Changes in transport operating costs, as well as fuel consumption, emissions and expected crashes for common urban logistics chains, need to be estimated.

Vehicle routing and scheduling procedures and traffic simulation models can be used to estimate the changes in vehicle kilometres of travel, average speeds and travel times. Optimisation models can be developed to estimate the costs for intermodal transport (road and rail) options.

Changes to operating costs, reliability, externality costs, and crash costs can be estimated based on predicted changes to the characteristics of the urban logistics chain networks, fleets and demand.

9.4.1 Time access restrictions

In many central city areas, it is popular to restrict delivery vehicles to specific times (*e.g.*, mornings) to improve liveability through less vehicle pedestrian conflicts, emissions, noise, and visual intrusion. There is a need to consider the impacts of time access restrictions on the vehicle routes of carriers as well as the environment.

Administrators must determine the times when freight vehicles are not allowed to enter as well as the area where the restrictions will apply. Designing effective schemes for minimising social impacts involves analysing the temporal and spatial patterns of pedestrian flows and crashes.

The City of Melbourne has developed a Pedestrian Counting System that records and displays pedestrian count data at various locations in the city (City of Melbourne – Pedestrian Counting System). This type of system can be used to identify areas and periods where there are high pedestrian volumes that can be used to determine access restrictions for freight vehicles. A data visualisation tool allows patterns to be animated. Historical data can be used to analyse and predict pedestrian flows (Pfiester & Thompson, 2021).

In addition to analysing the social benefits of time access restrictions, the effects on productivity of carriers should also be considered. Carriers have a high degree of heterogeneity that can have a major effect on the operation of urban freight vehicles. Many trucks and vans are operated by retailers and producers of goods with transport being ancillary to their main business. There are also a substantial proportion of for-hire freight transport and logistics companies that can have a wide variety of geography coverage (local or international).

Fleet optimisation procedures were developed to predict the effects of impacts of increasing the duration and size of the restricted zone in Seville. An increase in total distance travelled and transport costs for freight transport companies were found to increase exponentially with intensity of access time window restrictions measured by the zone size and length of restriction (Muñuzuri et al., 2013). More vehicles are also required to perform deliveries when time access restrictions were present. Quak and de Koster (2009) also estimated that the cost impact of time windows was largest for retailers who combine numerous deliveries in one vehicle round-trip.

9.4.2 Vehicle restrictions

It is common for there to be restrictions for certain classes of freight vehicles that use traffic links in metropolitan areas. Infrastructure constraints, safety and aesthetics can influence the permitted length, width, height and weight of freight vehicles.

The effects of prohibiting large freight vehicles using the inner area of Sao Paulo from 5 am to 9 pm on weekdays and between 10 am and 2 pm on Saturdays, have been investigated (Zambuzi et al., 2016). This scheme has resulted in a substantial growth in smaller trucks that are more agile but are less productive. An optimisation model was developed to investigate the effects of this prohibition scheme considering range of customer densities, drop sizes and distribution centre locations. Generally larger vehicles were estimated to have lower emissions. The model predicted that marginal increases in the permitted size of freight vehicles, as well as greener vehicles, would lead to reduced air pollution and the total number of vehicles used.

Due to vehicle restrictions carriers are typically forced to use smaller vehicles instead of larger trucks. This often leads to more vehicles and additional distance travelled by freight vehicles resulting in higher operating costs for carriers and more environmental costs. Smaller, more frequent deliveries require less storage at establishments but are more disruptive for receivers.

Operating larger freight vehicles operating in urban areas have been estimated to lead to substantial benefits for carriers in terms of productivity as well as environment and safety (Thompson & Hassall, 2014; Hassall et al., 2019). Modelling of parcel deliveries and shipping container networks showed significant reduction in the number of trips, kilometres travel and operating costs when larger vehicles are used.

9.4.3 Low-emission zones/environmental zones (engine restrictions)

Zero-emission zones (ZEZ) have already been implemented in some European cities with many currently being planned. A common approach is to tighten or increase the access requirements for existing Low-Emission Zones (LEZ). This can be done by reducing time windows for non-EVs, introducing permits or removing parking and road space for non-EVs. There is a need to estimate the reduction in emissions from LEZs as well as the additional fleet costs.

London's LEZ has led to an increase in the replacement rate for older vehicles and marginal improvements in air quality (Ellison et al., 2013). The evaluation illustrates that a number of effects need to be considered in addition to air quality levels (PM_{10} and NO_X concentrations) including the composition of vehicles (vans, rigid and articulated trucks) operating within the LEZ and other areas, the turnover of fleets (new registrations).

The Intelligent Access Programme (IAP) in Australia (Intelligent Access Programme) utilises GPS to monitor heavy vehicles' compliance levels with approved access conditions allowing transport operators access roads to suit their business and operational needs. IAP increases governments confidence that freight vehicles are complying with agreed access conditions.

IAP is a voluntary programme that allows access or improved access to the road network in return for passive compliance monitoring. It facilitates the right vehicle, being on the right road at the right time. Carriers install certified telematics devices that enable the compliance behaviour of operators to be assessed. IAP Service Providers (IAP-SPs) are private sector monitoring companies certified to provide telematics services (*e.g.*, hardware, software, and associated processes). In addition to Participant Reports (PRs), IAP-SPs prepare Non-Compliance Reports (NCRs) for Australian Road Authorities. These do not necessarily lead to penalties.

IAP can improve productivity and road safety as well as reduce infrastructure wear and environmental and social impacts. It allows compliance with truck curfews and restrictions on specific roads to be cost effectively monitored. IAP allows compliance checking for freight vehicles operating under the Performance-Based Standards (PBS) scheme to have access or less restrictive access to the road network.

9.4.4 Vehicle load factor controls

Access to some city centres is restricted to freight vehicles with a high level of load capacity utilisation. Administrators need to determine the required vehicle utilisation, the area boundaries as well as how this will be enforced. The effect on emissions and congestion levels needs to be estimated.

9.5 TOLLS

Due to the growth of private sector investment in road infrastructure through Public Private Partnerships (PPPs), there has been an increasing number of toll roads constructed in many cities. Avoidance of urban toll roads by freight vehicles generally leads to increased social and environmental impacts such as safety, noise and air quality. There are a number of different types of tolling schemes that can be used to charge freight vehicles for using toll roads. Procedures need to be developed for determining the best type of charging scheme as well as the optimal charging levels for freight vehicles (Perera & Thompson, 2020a).

There is a need to estimate the impacts on key stakeholders relating to trucks and tolls in urban areas including government, toll companies, carriers, shippers, receivers and residents. Models need to be developed to estimate the effect of toll levels on the performance criteria of each stakeholder.

To determine the optimal toll levels for passenger and freight vehicles involves considering the objectives of the key stakeholders. Carriers aim to minimise their operating costs (including tolls). Toll companies aim to maximise their revenue or internal rate of return on investment from tolls. Residents wish to minimise noise impacts and emissions. Governments aim to minimise infrastructure, congestion levels, and crashes.

There is a substantial variation in toll charges ($/km) for trucks using various sections of Melbourne's CityLink toll road, with generally lower charges ($/km) for using multiple sections (Perera et al., 2016).

There is a need to estimate various performance measures for key stakeholders regarding tolls and urban freight. A bi-level modelling and multiple-objective approach was used to develop models for optimal toll levels for freight vehicles (Perera et al., 2020; Perera et al. 2021). The upper-level problem consists of a set of objectives for individual stakeholders. In the lower level, traffic assignment is performed for passenger and freight vehicles incorporating toll levels.

The Pareto front illustrating the trade-offs between user and environmental costs as well as revenue were estimated for various toll-pricing schemes. The Pareto front allow solutions to be identified that suite specific decision makers. Three objectives can be considered, the IRR value to better understand the acceptability of such schemes from an investor's point of view.

An assessment of the impacts of the Port Authority of New York and New Jersey's time-of-day pricing initiative on the behaviour of commercial carriers revealed that many carriers responded in multiple ways combining productivity increases and cost transfers and changes in facility usage (Holguín-Veras et al., 2006). This study highlights the need to understand more about the balance of power between carriers and receivers. Although carriers can benefit from operating during off-peak periods, this is only possible if their customers are willing to work at these times.

9.6 ROAD PRICING

Road pricing schemes aim to reduce congestion and emissions. They can be designed to discriminate against older vehicles. Possible side effects can be extra kilometres travelled by freight vehicles and peak avoidance in the case of congestion charging. It is important to investigate the different reactions of road pricing schemes by for-hire carriers and private carriers (Quak & van Duin, 2010).

A multi-agent systems model incorporating vehicle routing with time windows, electronic markets and learning to evaluate both cordon-based road pricing for freight vehicles for B2C deliveries was developed by Teo et al. (2012). Cordon-based was found to reduce pollution more than distance-based pricing but had less impact on areas outside the city.

A multi-agent system (MAS) model incorporating reinforcement learning was developed to evaluate the short-term impact of distance-based road pricing on carriers, shippers, administrators and customers in Osaka (Teo et al., 2014). A load factor control scheme implemented as a joint scheme with the urban freight road pricing was also considered. A range of performance measures were predicted including carriers profit, carriers costs, shippers costs, distance travelled by trucks, number of trucks used, number of customer complaints, and emissions (nitrogen oxide, NOx, carbon dioxide, CO_2, and SPM). The model predicted that the City Logistics joint scheme has the potential of improving average daily load factors and reduce emissions in comparison with the current situation.

9.7 CONCLUSIONS

It has become common for governments to introduce access restrictions to improve air quality, noise and vibration levels in specific areas within cities. Analytics are required to predict the benefits of access restrictions to ensure that policy objectives are achieved.

The provision of adequate facilities for freight vehicles to unload and load goods is important for ensuring carrier efficiency and traffic network and amenity levels. Models are required to determine the best size and location of loading areas.

Loading dock booking systems can reduce delays for carriers and improve traffic flow as well as provide information on freight generation. It is also important to have analytical procedures that can determine the best toll levels for urban freight vehicles as well as identify optimal road pricing schemes.

REFERENCES

Aiura, N. and Taniguchi, E. (2005). Planning on-street loading-unloading spaces considering the behaviour of pickup-delivery vehicles, *Journal of the Eastern Asia Society for Transportation Studies*, 6, 2963–2974.

Alho, A.R., De Abreu, E.S., Joao, de S., Jorge, P. and Blanco, E. (2018). Improving mobility by optimizing the number, location and usage of loading/unloading bays for urban freight vehicles, *Transportation Research Part D*, **61**, 3–18.

Aljohani, K., Stokoe, M., Tinsley, N. and Thompson, R.G. (2021). Predicting freight demand for planning loading docks, *Proceedings Australian Transport Research Forum (ATRF)*, Brisbane.

Browne, M., Allen, J., Nemoto, T., Visser, J. and Wild, D. (2008). City access restrictions and the implications for goods deliveries, In. Taniguchi, E. and Thompson, R.G. (eds.), *Innovations in City Logistics*, Nova Science Publishers, New York, 17–36.

Chiara, D.G. and Cheah, L. (2017). Data stories from urban loading bays, *European Transport Research Review*, **9**, 1–17.

Chloupek, A. (2012). i-Ladezone: Intelligent monitoring & routing of loading zones in Vienna, *Proceedings 19th ITS World Congress*, Vienna.

Ellison, R.B., Greaves, S.P., Hensher, D.A. (2013). Five years of London's low emission zone: Effects on vehicle fleet composition and air quality. *Transportation Research Part D*, **23**, 25–33.

Hassall, K., Thompson, R.G. and Cowell, K. (2019). The evolution of a high productivity urban e-commerce delivery vehicle using Australian Performance Based Standards, *Logistics and Transport*, **43**, 73–80.

Holguín-Veras, J., Wang, Q., Xu, N., Ozbay, K., Cetin, M., Polimeni, J. (2006). The impacts of time of day pricing on the behaviour of freight carriers in a congested urban area: Implications to road pricing, *Transportation Research Part A*, **40**, 744–766.

Iwan, S., Kijewska, K., Johansen, B.G., Edihammer, O., Malecki, K., Konicki, W. and Thompson, R. G. (2018). Analysis of the environmental impacts of unloading bays based on cellular automata simulation, *Transportation Research*, *Part D*, **61**, A, 104–117.

Kijewska, K., Iwan, S., Nurnberg, M., Małecki, K. (2018). Telematics tools as the support for unloading bays utilization. *Archives of Transport System Telematics*, **11**, 23–28.

Letnik, T., Farina, A., Mencinger, M., Lupi, M. and Bozicnik, S. (2018). Dynamic management of loading bays for energy efficient urban freight deliveries, *Energy*, **159**, 916–928.

McLeod, F. and Cherrett, T. (2011). Loading bay booking and control for urban freight, *International Journal of Logistics Research* Applications **14**, 385–397.

Muñuzuri, J., R. Grosso, P. Cortés and Guadix, J. (2013). Estimating the extra costs imposed on delivery vehicles using access time windows in a city, *Computers, Environment and Urban Systems*, **41**, 262–275.

Perera, L., Thompson, R.G. and Yang, Y. (2016). Analysis of toll charges for freight vehicles in Melbourne, Proceedings 38th Australian Transport Research Forum (ATRF), November 16–18, 2016, Melbourne.

Perera, L., Thompson, R.G. and Wu, W. (2021). Determining optimum toll charges for freight vehicles considering multi-stakeholder objectives in urban conditions, *Transportation Journal*, **60**, 171–207.

Perera, L., Thompson, R.G. and Wu, W. (2020). A multi-class toll-based approach to reduce total emissions on roads for sustainable urban transportation, *Sustainable Cities and Society*, **63**, 102435.

Perera, L. and Thompson, R.G. (2020). Road user charging for urban freight vehicles: A systems approach, *Journal of Transportation Technologies*, **10**, 214–243.

Quak, H. J. and de Koster, M. B. M. (2009). Delivering Goods in Urban Areas: How to Deal with Urban Policy Restrictions and the Environment, *Transportation Science*, **43**, 2, 211–227.

Quak, H. and van Duin, R. (2010). The influence of road pricing on physical distribution in urban areas, *Procedia Social and Behavioral Sciences*, **2**, 6141–6153.

Quak, H. (2015). Access restrictions and local authorities' city logistics regulation in urban areas, Chapter 12, In: *City Logistics – Mapping the Future*, CRC Press, Taylor and Francis, Boca Raton, FL.

Sanders, D., Hancock, S. and Thompson, R.G. (2016). Managing city logistics with MobileDOCK, Paper AN-CP0324, *23rd ITS World Congress*, Melbourne, October 10–14, 2016.

Stokoe, M. (2017). Development of a courier hub in Sydney CBD, Proceedings 10th International City Logistics Conference, June 14–16, 2017, Institute for City Logistics, Phuket, 469–482.

Stokoe, M. (2019). Space for Freight – Managing capacity for freight in Sydney – A CBD undergoing transformation, *Transportation Research Procedia*, **39**, 488–501.

Teo, J.S., Taniguchi, E. and Qureshi, A. G. (2012). Evaluation of distance-based and cordon-based urban freight road pricing in e-commerce environment with multiagent model, *Transportation Research Record*, **2269**, 127–134.

Teo, J., Taniguchi, E. and Qureshi, A. G. (2014). Evaluation of load factor control and urban freight road pricing joint schemes with multi-agent systems learning models, *Procedia - Social and Behavioral Sciences*, **125**, 62–74.

Thompson, R.G. and Hassall, K. (2014). Implementing high productivity freight vehicles in urban areas, *Procedia - Social and Behavioral Sciences*, **151**, 318–332.

Thompson, R.G. and Flores, G. (2016). Understanding deliveries to towers in Melbourne, *Transportation Research Procedia*, **16C**, 510–516.

Thompson, R.G. and Zhang, L. (2018). Optimising courier routes in central city areas, *Transportation Research, Part C*, **93**, 1–12.

Zambuzi, N. C., Cunha, C. B., Blanco, E. Yoshizaki, H. and Carvalho, C. D. (2016). An evaluation of environmental impacts of different truck sizes in last mile distribution in the city of São Paulo, Brazil, Proceedings *6th International Conference on Information Systems, Logistics and Supply Chain, ILS Conference 2016,* Bordeaux.

Zhang, L. and Thompson, R.G. (2019). Understanding the benefits and limitations of occupancy information systems for couriers, *Transportation Research, Part C*, **105**, 520–535.

Zhang, J. and Thompson, R.G. (2021). Optimising Product Swaps in Urban Retail Networks, Proceedings, IPIC2021, *Proceedings, International Physical Internet Conference*, Shenzhen.

Chapter 10

Environmental sustainability

10.1 INTRODUCTION

Enhancing environmental sustainability is an important goal of city logistics. Environmental impacts of urban freight transport such as air pollution, noise and vibration can be reduced when city logistics solutions are implemented. To achieve more sustainable urban freight transport, collaboration between stakeholders is needed.

Cleophas et al. (2019) discussed collaborative urban freight systems, identifying two key types of collaboration, vertical and horizontal. Vertical collaboration is referred to line-haul and last-mile or multi-echelon distribution, whereas in horizontal collaboration partners serve the same or at least over-lapping parts of the transport network. They considered strategic, tactical, and operational planning horizons as well as different degrees of information exchange and planning centralisation.

Taniguchi et al. (2018) presented concepts of an integrated platform for innovative city logistics with urban consolidation centres (UCC) and transhipment points (TP). The concepts encompass joint delivery systems with shared use of pickup-delivery trucks, urban consolidation centres and transhipment points based on big data analytics. These systems involve public-private partnerships among shippers, freight carriers, UCC operators, municipalities, and regional planning organisations since the location of UCC and TP are highly related to urban land use plans and infrastructure provision as well as urban traffic management. ITS- and ICT-based communication systems for stakeholders are also essential for the efficient management of integrated platforms for both real-time and long-term operations. Integrated platforms can provide benefits such as improving the efficiency of urban deliveries and reducing CO_2 footprints.

DOI: 10.1201/9781003261513-12

10.2 JOINT DELIVERY SYSTEMS WITH URBAN CONSOLIDATION CENTRES (UCC)

Joint delivery systems with urban consolidation centres (UCC) (Allen et al., 2012) are a way to consolidate the flow of deliveries in the last mile. UCC are central elements in vertical collaboration that make it possible to group goods that have been addressed for certain customers into a single flow (Taniguchi & van der Hijden, 2000). These centres facilitate last-mile delivery by providing equipment that allows large vehicles to enter these areas, consolidating the flows that can be formatted to suit a delivery mode more adapted for the last mile.

Carriers are independent companies that receive transport orders from suppliers of the final customers. They deliver goods from their depot to a given final destination and use intermediate logistics platforms such as a UCC to optimise their own transport performance. Collaboration using common terminals and vehicles and even information systems among competitive freight carriers are needed to successfully operate UCC. UCC are usually located in the periphery area of cities and since the number of vehicles can be reduced from UCC to customers, they alleviate traffic congestion and decrease negative environmental impacts.

Joint delivery systems combined with UCC provide a good method for efficient and environmentally friendly urban freight transport systems. However, it is a challenge to successfully implement joint delivery systems with UCC, since there are uncertainties relating to the collaboration between competitive companies in terms of reduced logistics costs, service levels to customers and confidentiality of business information. Appropriate business models are required to understand the liability of UCCs (van Duin et al. 2016; Bjorklund et al., 2017). Socio-economic evaluation of UCC is critical for implementing and operating UCC (Isa et al., 2021; Aljohani & Thompson, 2021).

Regarding collaborations between stakeholders, joint delivery systems using urban consolidation centres (UCC) have been implemented in Tokyo. Figure 10.1 shows an example of a UCC in Tokyo for the multi-tenant buildings (MTB), Tokyo Sky Tree Town, which started in 2012. In this case three UCC are set for delivering goods to the MTB where shops, restaurants, offices, schools, and aquarium are located. Multiple freight carriers drop off their goods at UCC and a single freight carrier delivers goods to the MTB and final destinations of shops, restaurants *etc.* within the MTB. This freight carrier was involved in designing goods distribution systems within the MTB including the loading/unloading space, elevators, and information systems from the beginning of planning the MTB which was helpful for smooth and effective operations of logistics activities.

Figure 10.2 illustrates the collaboration between stakeholders who are participating in the joint delivery systems with the UCC. The developer

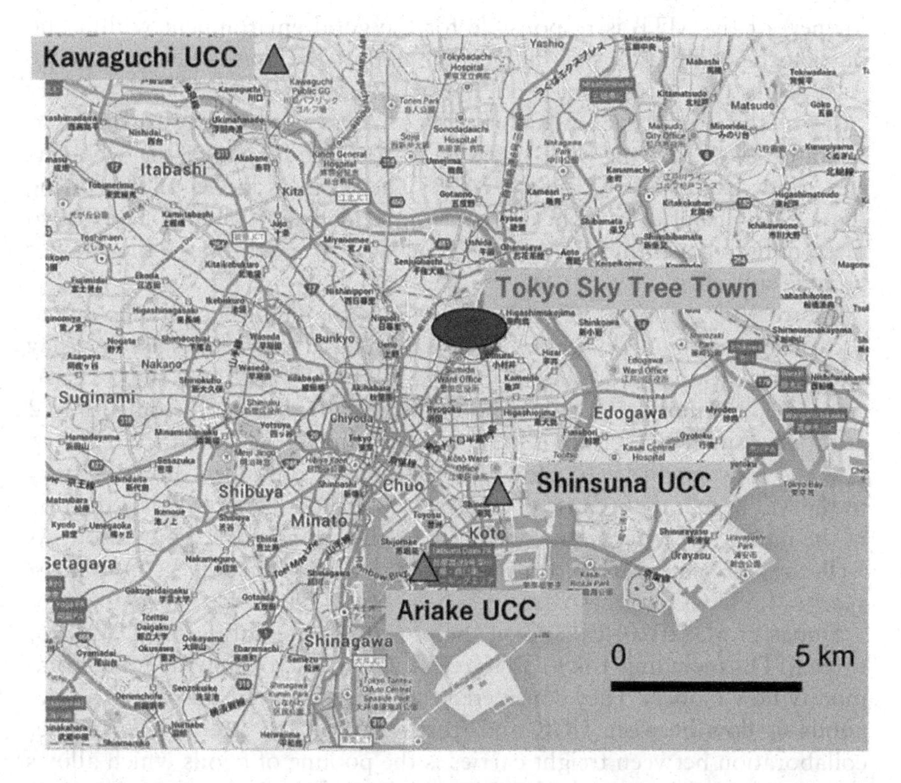

Figure 10.1 Multi-tenant building (Tokyo Sky Tree Town) and three UCCs in Tokyo.

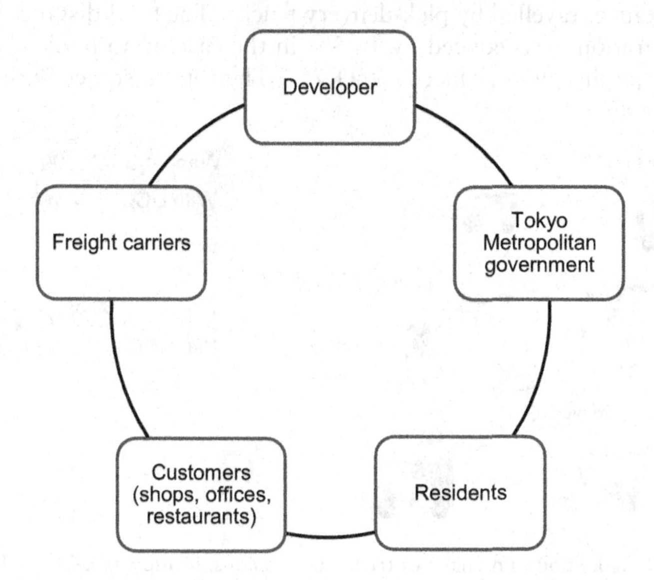

Figure 10.2 Collaboration of stakeholders for the joint delivery systems.

(owner) of the MTB is responsible for the good environment within and outside the MTB and they financially support the joint delivery systems. As a result, the system achieved a lower UCC fee by 33%–58% than other cases in the Tokyo area, which freight carriers need to pay to the UCC operator. Note that the UCC operator is the same freight carrier who carries goods from the UCC to the MTB. The business model of supporting the UCC is helpful for continuing the joint delivery systems. The Tokyo metropolitan government gives advice on legal issues, although they do not provide subsidies.

Figure 10.3 demonstrates the reduction in the number of trucks due to the consolidation of goods at UCC. The total number of trucks used per day with the UCC was reduced to 337 from 800 (by 58%) without the UCC. At each UCC the number of trucks was substantially decreased from 140 to 2 at the Shinsuna UCC, from 230 to 3 at the Ariake UCC, and from 100 to 2 at the Kawaguchi UCC. However, 330 trucks continued direct delivery to the MTB, mainly because of requests from shop owners. This reduction in the number of trucks can contribute to alleviating traffic congestion. As well, the CO_2 emissions were also reduced by 22% from 3,273 CO_2 tonnes/year to 2,532 CO_2 tonnes/year.

Dupas et al. (2020) calculated the effects of integrating four UCC for three MTBs including Tokyo Sky Tree Town in Tokyo applying multi-commodity flow models. They considered the integration of UCC which allows goods to flow between UCC and from any UCC to any MTB. Another collaboration between freight carries is the pooling of goods which allows them to use the same trucks for delivering parcels to the same destinations. They showed that both pooling and integration are effective in reducing the total distance travelled by pick-delivery trucks. The total distance with the full integration was reduced by 18.3% in the case of no pooling and that with the pooling was reduced by 11.7%. The total distance with pooling

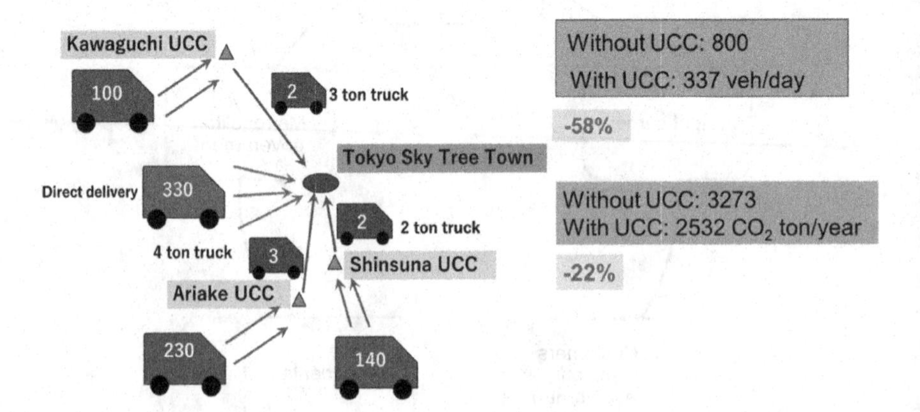

Figure 10.3 Reduction of number of trucks by the consolidation at UCC to Tokyo Sky Tree Town.

and full integration was reduced by 50.1% compared to the case without pooling and integration. Therefore, the collaboration between freight carriers by sharing the capacity of trucks and UCC are effective to decrease the total distance travelled which leads to cost savings and reduced environmental impacts. They also focused on CO_2, NOx and SPM. The integrated version of the network allows a reduction of CO_2 compared to the nonintegrated version; this reduction varies from 5% to 36% in the homogeneous fleet case.

10.3 CARGO BIKES

Cargo bikes or electric assisted cargo bikes have been used for sustainable last-mile deliveries in urban areas since they do not emit Green House Gas (GHG) and local Nitrogen Oxide (NOx) and Particulate Matter (PM). Cargo bikes are quieter and require less space for parking compared to conventional delivery trucks. However, the loading capacity of cargo bikes is lower than delivery trucks and their range is limited. Therefore, some micro depots or satellite facilities are needed to transship goods from delivery trucks or vans to cargo bikes. Evaluation of cargo bikes in terms of emissions needs to include the emissions by the feeder delivery trucks or vans to micro depots.

Llorca and Moeckel (2021) assessed the potential of cargo bikes and electric vans using an agent-based simulation model in Munich. They concluded that cargo bikes and electric vans are able to reduce CO_2 emissions, even after accounting for the emissions relating to the electricity production. With the low share of cargo bikes, the total distance travelled increased, since the reduction of van tours cannot compensate for the additional feeder trips from the distribution centre to the micro depots. Assmann et al. (2020) studied the implementation of cargo bike transshipment points in urban areas using a scenario-based model. Their findings demonstrated that using cargo bikes in urban deliveries could reduce GHG, PM10, and NOx emissions significantly. However, the choice of vehicles completing inbound and outbound processes and the strategies for siting urban transshipment points display widely differing and even conflicting potential to reduce emissions.

Sheth et al. (2019) analysed the costs of electric assist cargo bikes and delivery trucks in Seattle and pointed out that electric assisted cargo bikes are more cost-effective than delivery trucks in close proximity to the distribution centre: less than 2 miles (3.2 km) in the observed delivery route and 50 parcels per stop.

Elbert and Friedrich (2020) applied a hybrid agent-based and discrete event simulation model in Frankfurt to analyse the integration of cargo bikes into urban consolidation centre concept across multiple logistics service providers. Their simulation results showed that cargo bikes can be a suitable addition to urban consolidation concepts from both environmental and financial point of views.

10.4 CONCLUSIONS

Enhancing environmental sustainability is an important goal of city logistics. Joint delivery systems in urban areas with urban consolidation centres (UCC) are a way to consolidate the flow of deliveries in the last mile. Based on the case studies in Tokyo, the pooling of goods and the integration of UCC is effective to reduce the distance travelled by delivery trucks as well as the environmental impacts. Cargo bikes can be used for the last-mile delivery, but some micro depots or satellite facilities are needed to transship goods from delivery trucks or vans to cargo bikes.

REFERENCES

Aljohani, K. and Thompson, R.G. (2021). Profitability of freight consolidation facilities: A detailed cost analysis based on theoretical modelling, *Research in Transportation Economics*, 90, 101122.

Allen, J., Browne, M., Woodburn, A., and Leonardi, J. (2012). The role of urban consolidation centres in sustainable freight transport, *Transport Reviews*, **32**, 473–490.

Assmann, T., Lang, S. Müller, F., and Schenk, M. (2020). Impact assessment model for the implementation of cargo bike transshipment points in urban districts, *Sustainability*, 12, 4082.

Bjorklund, M., Abrahamsson, M., and Johansson, H. (2017). Critical factors for viable business models for urban consolidation centres, *Research in Transportation Economics*, 64, 36–47.

Cleophas, C., Cottrill, C., Jan Fabian Ehmke, J. F., and Tierney, K. (2019). Collaborative urban transportation: Recent advances in theory and practice, *European Journal of Operational Research*, 273, 801–816.

Dupas, R., Taniguchi, E., Deschamps, J.-C., and Qureshi, A. (2020). A multi-commodity network flow model for sustainable performance evaluation in city logistics: Application to the distribution of multi-tenant buildings in Tokyo, *Sustainability*, 12, 2180.

Elbert, R. and Friedrich, C. (2020). Urban consolidation and cargo bikes: A simulation study, *Transportation Research Procedia*, 48, 439–451.

Isa, S. S., Lima Jr., O. F., and Vieira, J.G.V. (2021). Urban consolidation centers: Impact analysis by stakeholder, *Research in Transportation Economics*, 90, 101045.

Llorca, C. and Moeckel, R. (2021). Assessment of the potential of cargo bikes and electrification for last-mile parcel delivery by means of simulation of urban freight flows, *European Transport Research Review*, 13, 33.

Sheth, M., Butrina, P., Goodchild, A., and McCormack, E. (2019). Measuring delivery route cost trade-offs between electric-assist cargo bicycles and delivery trucks in dense urban areas, *European Transport Research Review*, 11, 11.

Taniguchi, E., Dupas, R, Deschanmps, J-C., and Qureshi, A. G. (2018). Concepts of an integrated platform for innovative city logistics with urban consolidation centers and transshipment points. In: Taniguchi, E. and Thompson, R.G. (eds), *City Logistics 3: Towards Sustainable and Liveable Cities*, ISTE, London, 129–146.

Taniguchi, E. and van der Heijden, R.E.C.M., (2000). An evaluation methodology for city logistics, *Transport Reviews*, 20, 65–90.

van Duin, J.H.R., van Dam, T. Wiegmans, B., and Tavasszy L.A. (2016). Understanding financial viability of urban consolidation centres: Regent Street (London), Bristol/Bath & Nijmegen, *Transportation Research Procedia*, 16, 61–80.

Chapter 11

Disruption of networks

11.1 INTRODUCTION

The efficiency of commercial logistics depends on a complex mix of various factors such as demand, supply, transportation network, and availability of human and other resources. Under normal conditions, these factors remain stable, well predictable, reliable and scalable to some extent. Decision making thus becomes possible to be streamlined and pursue most common objectives of cost minimisation or profit maximisation along with customers' satisfaction. Natural or manmade disasters create disruptions in commercial logistics networks and require special response. In addition to the seasonal and somewhat forecastable natural disasters such as hurricanes, floods and droughts, logistics networks have suffered from a recent increasing trend of sudden disruptions due to mega-scale disasters such as earthquakes (*e.g.*, East Japan Earthquake, 2011) and pandemics such as COVID-19. Each disaster can impact multiple factors mentioned earlier, for example, COVID-19 caused an increase in e-Commerce demand. Although transportation networks were not disturbed many companies experienced a shortage of drivers. New challenges such as contactless deliveries also arose during the COVID-19 disruption.

Unexpected natural disasters such as earthquakes severely impact almost all sectors of logistics. As people seek refuge in shelters, the location and amount of demand becomes dynamic and unpredictable. Transportation networks are disturbed and supplies becoming insufficient and only intermittently available resulting in a chaotic decision making environment emerging at least in the initial response period. A shortage of human resources (such as drivers) and trucks further complicates the situation (Holguín-Veras et al., 2014). Maximisation of the coverage and minimisation of the delay in the response overtakes the usual objective of cost minimisation. The focus changes from commercial logistics to humanitarian logistics.

DOI: 10.1201/9781003261513-13

11.2 HUMANITARIAN LOGISTICS IN DISASTERS

Most humanitarian logistics research focuses on the last mile delivery of relief goods to the shelters, which is considered the most difficult and the critical part of the whole humanitarian supply chain. Barbarosoğlu & Arda et al. (2004) defined disaster relief operations as the delivery problem consisting of transportation of first aid material, food, equipment, and rescue personnel from disaster management centres to shelters, which are geographically scattered over the disaster region. They included safe and rapid evacuation of people affected by the disaster as an important part of relief operations. As mentioned earlier, the dynamic nature of the demand and their location, partially available networks and slow recovery and intermittent supplies are the main factors responsible for the limited academic research on this complex problem (Sheu, 2007). Most of the work appears in practitioners' magazines after major disasters sharing specific experiences related to them (Kovacs & Spens, 2007). In academic research, last mile disaster relief operations have been modelled either as a routing problem of relief goods, or a facility location problem of distribution centres or as a combination of facility location and routing problem.

11.2.1 Routing problem

The last leg of humanitarian logistics consists of distributing relief goods from distribution centres to shelters. This can be modelled as a vehicle routing problem with peculiar objective functions and constraints attributed to the humanitarian logistics. For example, Balcik et al. (2008) used a two-phase approach to optimise last mile relief delivery. In the first phase, all possible combinations of demand locations were considered and in the second phase, the route with the minimum travel time was found by solving the Traveling Salesman Problem (TSP) for each combination. Holguin-Veras et al. (2012) described that the periodicity of the regular humanitarian logistics is relatively repetitive in nature. Such a nature of relief distribution operation requires a multi-period evaluation. Lin et al. (2011) addressed the last mile delivery with a multi-item, multi-period relief distribution model considering time windows and split delivery concepts. They also introduced priority of items and a penalty if the item is not delivered on time. They also considered the difference between the service levels of two demand points.

It is a well-known fact that the resilience of human beings goes on decreasing with sustained lack of food, water and medicine. However, the demand for food, water and medicine is not additive in itself, *i.e.*, shortage of one day cannot be carried over to the next day but the resulting shortage does decrease the resilience of the affected person in the next day. Qureshi and Taniguchi (2017) presented a relief distribution model that explicitly incorporated the complex characteristics of relief demand that includes its non-additive nature. They introduced a dynamic penalty function that depends on the shortage of relief supply which goes back to the original value once the supply matches the demand. They presented a realistic case study based on data of a ward in Osaka, Japan. After the 2011 East Japan mega earthquake disaster, there has been a renewed concern over possibility of occurrence of a Tonankai Trough (fault line near Tokyo) or Nankai Trough (fault line near Kinki area, Osaka and Kochi) earthquake. Recently, the expected worst-case magnitude of such an earthquake has increased from M8.8 to M9.1. It is expected that Osaka may receive shocks of seismic intensity of lower 6 (Japanese seismic scale) due to the Nankai Trough earthquake. Considering that the relief distribution operation in such conditions can last for many days, Qureshi and Taniguchi (2017) used a multi-period vehicle routing problem with penalties on the shortage of relief goods. For each period, they used a double layer representation in their proposed genetic algorithm, where one layer represented the routing decision (*i.e.*, the customers' chain), whereas a synchronised layer contained the amount of delivery goods for the customers to evaluate the dynamic penalty function mentioned earlier. It was found that both routing cost and shortage penalty objectives actively take part in overall optimisation of the relief distribution.

Rath and Gutjahr (2014) considered three objective functions *viz*. minimisation of depot costs, budget requirements and maximisation of the supply at each shelter location. Vitoriano et al. (2011) also presented a multi-objective formulation that included equity and priority in the model formulation. Abounacer et al. (2014) added minimisation of the difference between the demand and the supply (*i.e.*, the shortage) along with minimisation of the number of emergency workers required. Yi and Kumar (2007) presented a multi-period relief distribution model, where the unmet demand in a period is prioritised (penalised) in the next periods.

11.2.2 Location-routing problem

Another stream of relief distribution-related research combines the optimal location of the relief facilities as well. Wang et al. (2014) addressed the last mile of humanitarian logistics as a location-routing problem. They considered multiple objectives which included minimisation of the cost (routing and distribution centres), minimisation of maximum route length as well as the maximisation of the minimum route reliability. Each link in their network was associated with a predefined probability of its availability. Ceselli et al. (2014) presented a column generation based algorithm to solve the vaccine delivery problem in the aftermath of a disaster or emergency that was formulated as a location-routing problem. Vahdani et al. (2018) further added the road repair objective to the set of multi-objectives.

Moshref-Javadi and Lee (2016) have addressed a latency location-routing problem (LLRP) that minimises the arrival time at the different shelter locations instead of cost.

Ponboon et al. (2015) presented an application of the location-routing problem with time windows on the relief distribution operation in Ishinomaki, Japan in the aftermath of the Great East Japan earthquake and Tsunami disaster of 2011. As many as 3,417 people were killed, and 535 people are still officially missing (as of January 2013) (Nakanishi et al., 2013). About 53,742 houses were affected, of which 22,357 houses were completely destroyed, 11,021 houses were severely damaged, and 20,364 houses were partially damaged (Taniguchi & Thompson, 2013). The official figures of the number of evacuees were reported on 11 April 2011, a month after the disaster, to be 30,930 people, who were distributed to 152 shelters (shown in Figure 11.1). The sports park located in the north of the city (depot 1 in Figure 11.1) was used by the Japan Self Defense Force (JSDF) and private freight carriers in the operation after the earthquake. The market located on the west of the city (depot 2 in Figure 11.1) was also used after closing down the sports park in September 2011 (Taniguchi & Thompson, 2013). Usually in humanitarian efforts, a number of volunteers are available to work without pay and in some cases, the expenses of relief operations are covered by the government or non-governmental

Figure 11.1 Location of 10 proposed depots in the city of Ishinomaki.

Table 11.1 Scenarios of location routing in Ishinomaki case study

Scenario	*Number of depot*	*Depot and vehicle cost*	*Time windows*
1 (Real Operation)	2	-	-
2	10	-	-
3	10	-	✓
4	10	✓	-
5	10	✓	✓

organisations. These actual costs are assumed to be implicitly added inside the real distribution operation. In order to access what changes in the relief planning occur if the depot opening cost, the vehicles, and the operating costs such as fuel were considered, various scenarios were created (given in Table 11.1) and were assessed by solving the corresponding location-routing problem.

Scenario 1 represents the real operation (however, costs are recalculated) where only two depots with unlimited capacities are operated located at the sports park (depot 1) and the market (depot 2). Furthermore, the vehicle cost and operating costs are not taken into account. In scenario 2, eight new depots are proposed (marked as depots 3–10 in Figure 11.1). Their potential sites were determined based on their location, available spaces, and essential facilities. The capacity of each depot is estimated by its type and size. Scenario 3 is based on the lessons learned from the disaster that considered the volunteers or staff from relief organisations having a variety of functions to manage on an hourly basis and also there may be desirable visiting times at shelters. Therefore, hypothetical time windows (a_i and b_i) were assigned to each shelter randomly. The average time window width was kept equal to 132 time units. Scenario 4, considers the operating costs (including the depot opening cost and vehicle cost), to represent a scenario where the available resources such as vehicles and staffs are limited. Finally, in scenario 5, both operating costs and time windows are considered simultaneously. For solving the underlying LRPTW instance for these scenarios, Ponboon et al. (2015) proposed a Tabu search solution approach. The results are presented in Table 11.2.

In scenarios 1 and 2, only the routing parts are calculated as they do not consider the depot opening cost. The LRPTW algorithm prefers to open as many depots as long as the capacity constraints are satisfied, *i.e.*, depot and vehicle capacities are not violated. As expected, since the number of depots and the number of vehicles are increased, a relatively lower value of average truckload and total distance was obtained. In such settings it is possible that each shelter can be assigned to the nearest depot. Therefore, truck routes become shorter. In scenario 3 when time windows were assigned, although the number of depots was reduced to only seven depots, the number of trucks and travelling distance both increased. This effect and phenomenon are unique for LRP and LRPTW as described in

Table 11.2 Results of the LRPTW application in humanitarian logistics in Ishinomaki

Scenario	*Number of potential depot*	*Depot and vehicle cost*	*Time windows*	*No. of open depot*	*Open depot*	*No. of truck at each depot*	*Total no. of truck*
1 (Real Operation)	2	-	-	2	1,2	13,3	16
2	10	-	-	8	2,3,4,5,6, 7,9,10	1,2,4,1,4,2,1,3	18
3	10	-	✓	7	1,4,6,7,8, 9,10	1,6,4,3,2,1,3	20
4	10	✓	-	3	1,3,8	6,6,3	15
5	10	✓	✓	3	1,3,8	7,5,3	15

Scenario	*Number of potential depot*	*Depot and vehicle cost*	*Time windows*	*Average truck load*	*Average distance (km)*	*Total distances (km)*
1 (Real Operation)	2	-	-	0.92	15.88	254.03
2	10	-	-	0.81	8.11	146.01
3	10	-	✓	0.73	9.17	183.38
4	10	✓	-	0.98	13.60	203.95
5	10	✓	✓	0.98	17.06	255.87

Ponboon et al. (2013) since route length (in terms of the number of shelters) becomes limited due to the time window constraints. Scenario 3 also results in the lowest average truckload. When the depot opening cost is added in scenario 4, the depot location part dominated the LRPTW calculation process. Here, the algorithm tries to open the least possible number of depots thus resulting in only three open depots. The vehicle cost also played its part which guided the LRPTW algorithm to reduce the number of trucks to 15, resulting in higher average truckload. Due to few depots, trucks need to drive a relatively longer distance to reach shelters which are relatively far from their assigned depots. Finally, the travelling distances are slightly increased when the time windows are considered again in scenario 5 (similar to scenario 3). It is interesting to note that the real operation (scenario 1) performs almost similar to scenario 5 in terms of distance travelled, which can be related to day-to-day explicit running cost of the operations in terms of fuel cost. However, if only the LRP is considered (scenario 4), mathematical model-based operations would result in better performance, even if the depot and vehicle costs are not actually paid (based on the assumptions of the real operation, *i.e.*, scenario 1).

11.3 RECOVERY PROCESS

11.3.1 Debris removal

Natural disasters result in a huge amount of debris and waste that requires enormous clearance efforts at a large cost. For example, US Federal Emergency Management Agency (FEMA) estimated a sum of US$ 4 billion for debris cleaning costs after hurricane Katrina. In the 2011 East Japan mega earthquake disaster 127,531 houses were completely collapsed whereas the number of partially collapsed houses was 274,036. The Ministry of the Environment, Japan estimated the total amount of disaster debris including Tsunami deposits as about 280,000,000 tons (Ministry of Environment, Japan, 2014).

Collection of debris and transportation to disposal sites forms the major portion of the total disaster management cost. It is a costly and complex operation, often extending over months or even years after the disaster. Debris collection and transportation operations can be considered similar to the winter gritting problem (Tagmouti et al., 2007) or to the reverse logistics problem (Cardoso et al., 2013). Nonetheless, removal of debris after a disaster presents unique challenges, for example, debris can completely block the access from one road section to the others, therefore, the sequence of visiting and servicing arcs becomes very important in an access clearing route. Similarly, the access clearing routes can only be carried out in a particular sequence, considering that one completed route will affect the access possibility for the other routes. Pramudita et al. (2012) documented this problem as a capacitated arc routing problem (CARP) with access possibility constraints and presented a Tabu search algorithm to solve this problem on a hypothetical, small scale test network. Later, a case study in the Tokyo area was presented by Pramudita et al. (2014) based on the simulation results of debris generation by Hirayama et al. (2010) in the wake of a predicted large-scale earthquake in the Kanto region.

Qureshi et al. (2016) presented a realistic case study using data from Rikuzentakata City, one of the worst-hit areas of the Great East Japan Earthquake. The number of deaths there was recorded as 1,599 people with 207 still missing. About 3,805 houses were completely damaged, 240 were severely damaged and 3,986 were partly damaged. The earthquake and Tsunami resulted in a total of 1,674,000 tons of disaster debris, which was collected and deposited in local (intermediate) depots (Fire Department Report, 2014). Qureshi et al. (2016) considered that the entire pre-disaster network, which consisted of about 488 arcs, is affected by the debris. Considering a depot located along the boundary of the network at the junction of two major national routes, they modelled the debris collection and transport problem with a capacitated arc routing problem (CARP) with access possibility constraints. Although such models assume the clearance and reclamation of all network arcs, in a later field visit by the authors, it was found that many of the roads shown in the network were still not restored and probably forfeited to create a buffer between the new developments and the ocean. Interviews with the disaster management teams can highlight such network arcs, which can then be excluded from the input data to create more practical modelling applications.

In the capacitated arc routing problem with access possibility constraints (Pramudita et al., 2012; Qureshi et al., 2016) the CARP network is transformed into a capacitated vehicle routing problem (CVRP) network with two nodes replacing each arc. As mentioned earlier, large disasters (such as earthquakes and Tsunami) create a debris problem where all arcs are to be serviced, and hence the access to arcs, away from the depot, is usually blocked by the nearer arcs. Both Pramudita et al. (2012) and Qureshi et al. (2016) used the concept of the access probability matrix, which is similar to the adjacency matrix, with entries of zero and one, where one represents opened access. During the solution process, if the debris collection and transportation vehicle starts the route from depot 1 and first up services the nearby arcs (connected to depot). This operation changes the access possibility matrix, which becomes more populated with entries of ones as the route progresses. Therefore, at the earlier stages of the routing solution, the possibilities for the exchange operator (which switches the position of arcs in typical Tabu search algorithms) become limited. Qureshi et al. (2016) utilised this condition and presented an improved granular Tabu search algorithm taking advantage of the access possibility constraint.

Cheng et al. (2018) formulated a reliability index for waste transport in the aftermath of a disaster. Arguing that both the waste generation and waste cleanup capacity are random variables, where waste generation depends on the type, severity and location of the disaster and waste cleanup capacity depends on the target waste cleanup time and the number of trucks available and their capacities, available infrastructure (number of routes) and the location and state of the waste management infrastructure (such as landfills). Formulating the probability function of the performance (*i.e.*, the difference between the waste cleanup capacity and the waste generation), Cheng et al. (2018) introduced the reliability index as the ratio of the mean of the performance to its standard deviation. They also provided a formulation to obtain the number of required vehicles to maximise the reliability index for a given set of data (such as maximum cleanup cost, landfill capacity and so on, which define the waste cleanup capacity) for a particular disaster. Results of their hypothetical disaster scenarios (of bush fires) suggest that a well optimised mix of the vehicle type can improve the performance of waste cleanup, whereas larger capacity vehicles are more efficient as compared to the smaller ones. In their later work (Cheng et al., 2019), they extended the concept of the failure probability (with negative performance function) to individual routes from a disaster area to the waste disposable sites and provided an algorithm (based on event tree analysis) which can analyse and predict the most likely failure modes in terms of chain of route failures. As their analysis identifies the most critical routes (both in terms of the most improving and the most degrading (bottleneck) routes), it provides valuable information for disaster managers to strengthen these routes in pre-disaster preparation times.

Concept of the transfer station, *i.e.*, intermediate facilities (providing space for storage, sorting, recycling, *etc.*) is quite well established in day-to-day municipal solid waste management systems. However, in cases of disasters, where the location and amount of the disaster waste generation are uncertain, early location of such temporary disaster waste management sites (TDWMS) is essential for faster and more efficient disaster waste cleanup operations. Cheng et al. (2022) considered the location of TDWMS and routing to and from them as a multi-period 2E-LRP. Trucks start from a given depot and collect waste from various points (assuming less than truck load waste lots for each time period) and first transfer it to TDWMS (whose number and location is to be optimised) creating the first echelon. The second echelon consisted of the truck routes between TDWMS and final disposal sites. A further decision complexity is added in the form of sub-cycles in the first echelon, where trucks cycle between collection nodes and TDWMSs without going back to the depot.

11.3.2 Network recovery

Road infrastructure often gets affected by large-scale natural disasters, especially bridges, even though cities have been focusing on strengthening their road networks. However, still there is a possibility that some road links may not be available or not working at their full capacity due to damage and/or debris resulting from a large-scale disaster. Research related to the network availability or recovery for conducting humanitarian logistics is limited due to the dynamic nature of the problem. As mentioned earlier, Wang et al. (2014) considered the availability of a link with an associated probability which was predefined for each link of their network. Their analysis, however, did not consider the recovery (improved probability) of the link. Qureshi and Taniguchi (2020) addressed this issue in a scenario analysis by considering the available capacity of the road network as well as the ideal location of the emergency response depot in a case study based on Minato ward of Osaka City, which may get seriously affected due to Nankai Trough earthquake and resulting Tsunami (as mentioned in Section 11.1). They considered the designated location of shelters in the Minato ward along with four possible choices of the depot in their case study. In ideal conditions, the ward office of the ward is considered as the depot but there exists a high probability that the Minato ward office may itself be affected by a Tsunami and thus becomes unavailable. Therefore, Qureshi and Taniguchi (2020) considered three other depots (at other Osaka City government buildings) located at relatively higher altitude, which are less likely to be affected by the Tsunami. They considered a GIS-based road network of Osaka City, which includes the travel times in daytime congested situation. This travel time was multiplied by a factor of two to represent travel times in the post-disaster situation. To analyse the effect of network availability, they considered four scenarios for each depot. Scenario (S1) represented a fully available road network, then based on the paths connecting all the shelters and depots in S1, critical bridge segments were determined (on a trial and error basis), which could cause big disruptions to the paths obtained in S1. In scenario (S2), all of these critical bridge segments were assumed to be unavailable. Two additional scenarios (S3 and S4) were created for each depot considering the restoration of one of the unavailable bridges of scenario (S2). They applied a multi-period relief distribution model to this series of scenarios. Their results found that if the depot is located right inside the affected area (*i.e.*, at the Minato ward office in this case), it gives the best results (in fact, S2 scenario doesn't change its reliability [in terms of routing from it to the shelters] and hence, S3 and S4 were not performed for it). However, if this depot is also assumed to be affected by the earthquake and Tsunami, other depots with multiple choices to reach the affected area performed better than those with only limited access options. Their analysis can also be used to identify the priority of critical infrastructure for restoration operations.

Before the occurrence of a disaster, it is assumed that the road network is working in equilibrium traffic flow conditions (Wardrop Equilibrium) where no driver can reduce his/her travel time by switching to another path between its origin and destination. This equilibrium condition is achieved based on various factors, which also include the capacity of the road links and travel volume on them. In a post-disaster scenario, both capacity and volume on the roads can be very different due to the damage to many links. Usually, travel time is used as the main indicator/objective in the post-disaster road network recovery problems. Kaviani et al. (2020) presented a road network recovery problem based on the network design framework that optimises the recovery process from not only travel time perspective but also considered the demand fulfilment of different classes of vehicles (such as freight vehicles).

11.4 LOGISTICS IN PANDEMIC TIMES (COVID-19)

Logistics under a large-scale pandemic situation (such as COVID-19) can also be seen as logistics with disrupted logistics networks. The routine of the logistics activities can be affected in pandemics due to the lack of drivers (being sick or quarantined) and the emphasis on contactless deliveries (Chen et al. 2021). As mentioned in Chapter 8 (Section 8.2.1), autonomous delivery robots were used in Wuhan, China to deliver supplies to hospitals and residents (Stone, 2021). In their analysis, along with the economic and efficiency analysis (presented in Section 8.3.3), Taniguchi and Qureshi (2021) also discussed the issue of contactless delivery in terms of an index based on the number of human-to-human contacts. Although delivery systems only using autonomous delivery vehicles (ADVs) (such as DDP shown in Figure 8.1c) present the least number of human-to-human contacts, Taniguchi and Qureshi (2021) argued that systems using mothership and ADVs (such as the FSTSP mothership) and VRP-D shown in Figure 8.1b and e, respectively, are better options as they also provide more economic and efficient delivery systems.

11.5 CONCLUSIONS

Concepts of commercial logistics such as vehicle routing problems, location-routing problems, reverse logistics, *etc.* can be adapted for better planning to cope with the logistics and transportation challenges posed by disasters and pandemics. These models already incorporate conditions such as limited availability of supplies, human resources and infrastructure availability, which are often associated with disaster response. The scope of such models is present at both pre-planning and post-disaster management stages. At pre-planning stage, these can be used to identify critical infrastructure links and help in improving the resilience by retrofits/strengthening such infrastructure links and by allocating and pre-positioning of the resources. At post-disaster stage, these models can help in efficient distribution of relief goods and in traffic management.

REFERENCES

Abounacer, R., Rekik, M. and Renaud, J. (2014). An exact solution approach for multi- objective location-transportation problem for disaster response, *Computers and Operations Research*, **41**, 83–93.

Balcik, B., Beamon, B.M. and Smilowitz, K. (2008). Last mile distribution in humanitarian relief, *Journal of Intelligent Transportation Systems: Technology, Planning, and Operations*, **12**, 51–63.

Barbarosoğlu, G. and Arda, Y. (2004). A two-stage stochastic programming framework for transportation planning in disaster response, *Journal of the Operational Research Society*, **55**, 43–53.

Cardoso, S.R., Povoa, B. and Relvas, S. (2013). Design and planning of supply chains with integration of reverse logistics activities under demand uncertainty, *European Journal of Operational Research*, **226**, 436–451.

Ceselli, A., Righini, G. and Tresoldi, E. (2014). Combined location and routing problems for drug distribution, *Discrete Applied Mathematics*, **165**, 130–145.

Chen, C., Demir, E., Huang, Y. and Qiu, R. (2021). The adoption of self-driving delivery robots in last mile logistics, *Transportation Research Part E*, **146**, 102214.

Cheng, C., Zhang, L. and Thompson, R.G. (2018). Reliability analysis for disaster waste management systems, *Waste Management*, **78**, 31–42.

Cheng, C., Zhang, L. and Thompson, R.G. (2019). Reliability analysis of road networks in disaster waste management, *Waste Management*, **84**, 383–393.

Cheng, C., Zhu, R., Costa, A.M., Thompson, R.G. and Huang, X. (2022). Multi-period two-echelon location routing problem for disaster waste clean-up, *Transportmetrica A: Transport Science*, **18**, 1053–1083.

Fire Department Report [Online in Japanese]. Available at http://www.fdma.go.jp/bn/higaihou/pdf/jishin/150.pdf (accessed March 1, 2015).

Hirayama, N., Shimaoka, T., Fujiwara, T., Okayama, T. and Kawata, T. (2010). *Establishment of Disaster Debris Management Based on Quantitative Estimation Using Natural Hazard Maps in Waste Management and the Environment V*, WIT Press. Available online at https://www.witpress.com/elibrary/wit-transactions-on-ecology-and-the-environment/140/21270 (accessed March 15, 2023).

Holguín-Veras, J., Jaller, M. and Wachtendrof, T. (2012). Comparative performance of alternative humanitarian logistic structures after the Port-au-Prince earthquake: ACEs, PIEs and CANs, *Transportation Research Part A: Policy and Practice*, **46**, 1623–1640.

Holguín-Veras, J., Taniguchi, E., Jaller, M., Aros-Vera, F., Ferreira, F. and Thompson, R.G. (2014). The Tohoku disasters: Chief lessons concerning the post disaster humanitarian logistics response and policy implications, *Transportation Research Part A: Policy and Practice*, **69**, 86–104.

Kaviani, A., Thompson, R.G., Rajabifard, A. and Sarvi, M. (2020). A model for multi-class road network recovery scheduling of regional road networks, *Transportation*, **47**, 109–143.

Kovacs, G. and Spens, K.M. (2007). Humanitarian logistics in disaster relief operations, *International Journal of Physical Distribution & Logistics Management*, **37**, 99–114.

Lin, Y.H., Batta, R., Rogerson, P.A., Blatt, A. and Flanigan, M. (2011). A logistics model for emergency supply of critical items in the aftermath of a disaster, *Socio-Economic Planning Sciences*, **45**, 132–145.

Ministry of the Environment, Japan. (2014). Progress on Treatment of Debris from the Great East Japan Earthquake (in Coastal Municipalities of the Three most affected Prefectures). Available at http://www.env.go.jp/en/recycle/eq/ptd20140326.pdf (accessed February 4, 2015).

Moshref-Javadi, M. and Lee, S. (2016). The latency location-routing problem, *European Journal of Operational Research*, **255**, 604–619.

Nakanishi, H., Matsuo, K. and Black, J. (2013). Transportation planning methodologies for post-disaster recovery in regional communities: The East Japan earthquake and tsunami 2011, *Journal of Transport Geography*, **31**, 181–191.

Ponboon, S., Qureshi, A.G. and Taniguchi, E. (2013). Integrated approach for location-routing problem using branch-and-price method, Hiroshima, JSCE 47th Conference on Infrastructure Planning and Management.

Ponboon, S., Qureshi, A.G. and Taniguchi, E. (2015). Location-routing problem for disaster relief operations, presented at the 6th International Symposium on Transportation Network Reliability, The Value of Reliability, Robustness and Resilience, August 2–3, 2015, Nara, Japan.

Pramudita A., Taniguchi, E. and Qureshi, A.G. (2012). Undirected capacitated arc routing problem in debris collection operations after disaster, *JSCE Journal Division D3 (Doboku Gakkai Ronbunshuu D3)*, **68**, 805–814.

Pramudita A., Taniguchi, E. and Qureshi, A.G. (2014). Location and routing problem of debris collection operation after disaster with realistic case study, *Procedia - Social and Behavioral Sciences*, **125**, 445–458.

Qureshi, A. G. and Taniguchi, E. (2017). A multi-period relief distribution model considering limited resources and decreasing resilience of affected population, *Journal of the Eastern Asia Society for Transportation Studies*, **12**, 57–73.

Qureshi, A. G. and Taniguchi, E. (2020). A multi-period humanitarian logistics model considering limited resources and network availability, *Transportation Research Procedia*, **46**, 212–219.

Qureshi, A. G., Taniguchi, E. and Yamamoto, M. (2016). A new Tabu search algorithm for collection and transport of the debris after disasters, 14th World Conference on Transport Research (WCTRS2016), 10–15 July, 2016, Shanghai, China.

Rath, S. and Gutjahr, W.J. (2014). A math-heuristic for the warehouse location-routing problem in disaster relief, *Computers & Operations Research*, **42**, 25–39.

Sheu, J.B. (2007). Challenges of emergency logistics management, *Transportation Research Part E*, **43**, 655–659.

Stone, T. (2021). Case study: China's JD.com launches fleet of 30 new autonomous delivery bots. Available at https://www.traffictechnologytoday.com/news/autonomous-vehicles/case-study-chinas-jd-com-launches-fleet-of-30-new-autonomous-delivery-bots.html (accessed March 4, 2022).

Tagmouti, M., Gendreau, M. and Potvin, J. (2007). Arc routing problem with time dependent service costs, *European Journal of Operational Research*, **181**, 30–39.

Taniguchi, E. and Qureshi, A.G. (2021). Emergency urban delivery management during the COVID-19 pandemic, report submitted to the Kinki Kensetsu Kyokai (in Japanese).

Taniguchi, E. and Thompson, R. G. (2013). Humanitarian logistics in the great Tohoku disasters 2011. In: Zeimpekis, V., Ichoua, S. and Minis, I. (eds.) *Humanitarian and Relief Logistics*. Springer, New York, 207–218.

Vahdani, B., Veysmoradi, D., Shekari, N. and Mousavi, S.M. (2018). Multi-objective, multi-period location-routing model to distribute relief after earthquake by considering emergency roadway repair, *Neural Computing and Applications*, **30**, 835–854.

Vitoriano, B., Ortuño, M.T., Tirado, G. and Montero, J. (2011). A multi-criteria optimization model for humanitarian aid distribution, *Journal of Global Optimization*, **51**, 189–208.

Wang, H., Du., L. and Ma, S. (2014). Multi-objective open location-routing model with split delivery for optimized relief distribution in post-earthquake, *Transportation Research Part E*, **69**, 160–179.

Yi, W. and Kumar, A. (2007). Ant colony optimization for disaster relief operations, *Transportation Research Part E*, **43**, 660–672.

Chapter 12

Future directions

12.1 INTRODUCTION

There is a need to develop enhanced analytics for planning, designing and operating future urban freight systems. Descriptive analytics constructed for understanding what happened in the past will need to be able to utilise big data collected from sensor networks that are described in Chapter 2. Improved techniques for detecting trends, patterns and incidents in real time are required. Predictive analytics aimed at determining what could happen in the future will need to incorporate more behavioural aspects and interactions between stakeholders. Gamification provides an effective means of collecting behavioural data. Agent-based modelling approaches incorporating learning agents using artificial intelligence described in Chapter 5 will need to be adapted to effectively evaluate initiatives and policies. Prescriptive analytics aimed at determining what should happen that are introduced in Chapter 4 will need to be more dynamic. Optimisation procedures will need to incorporate more uncertainty as well as interactions between multiple stakeholders.

Future analytics will need to be able to design and manage urban freight systems that are more sustainable and efficient when responding to changes in demand patterns from e-Commerce, ageing population and working from home. Improved analytics will be required to effectively design and evaluate open and shared concepts. New digital platforms will be necessary for facilitating collaborative freight systems.

Future analytics will be required to assist in the evaluation of initiatives for planning, designing, operating and managing urban freight systems that address the following key global challenges:

i. achieving net-zero emissions,
ii. enhancing public health: reducing emissions, noise, and road trauma,
iii. improving resilience: coping well with disrupted and congested networks from disasters and extreme weather, construction projects, special events, and incidents, and
iv. achieving the United Nations sustainable development goals.

DOI: 10.1201/9781003261513-14

This chapter describes a range of models that will be required to implement Hyperconnected City Logistics that has good potential for improving the sustainability and efficiency of urban freight systems. Several emerging technologies that have promise for addressing key challenges are introduced. Finally, an integrated platform for providing improved decision support is described.

12.2 HYPERCONNECTED CITY LOGISTICS

Hyperconnected City Logistics (HCL) based on the principles of the Physical Internet (Montreuil, 2011) and City Logistics (Taniguchi & Thompson, 2015) involves creating more sustainable urban distribution systems by promoting collaborative, integrated and open networks (Crainic & Montreuil, 2016). This involves developing new approaches to create multimodal networks, consisting of trucks, vans, bikes, and walkers with transfers of goods being conducted at micro-consolidation centres or parcel locker banks. Existing urban distribution networks can be transformed by shippers and carriers sharing vehicles and storage facilities to dramatically reduce distances travelled allowing electric and non-motorised vehicles to become more practical and viable.

A range of new analytical tools will be required to transform current urban freight systems that consist of largely closed and independent networks into open and shared networks. Procedures will be required by PI organisations for determining the right modes and load factors for the right loads, instead of the right mode for the right load. This will involve urban distribution systems typically utilising multiple modes, such as a combination of electric vans, bikes and trolleys. New methods for planning, designing and operating multi-modal routing and scheduling will be needed. Procedures for coordinating transfers between modes, and integrating vehicle movements for managing loading docks and bays will be necessary.

Improved analytics will be required for achieving the goals within the ALICE roadmap for implementing the Physical Internet by 2050 (ALICE, 2015). This will involve designing and assessing new opportunities and business models, creating efficient and automated distribution systems as well as sustainable and integrated urban logistics in the city mobility system. Information systems for designing innovative financial and legal layers will also be necessary.

Implementing HCL will require PI hubs for transferring goods between vehicles. As discussed in Chapter 9, booking or reservation systems at loading docks provide an effective means of reducing delays at major hubs. Due to the uncertainty of travel times in congested networks, improved coordination of docks and equipment for transferring goods between modes at hubs will be essential to ensure reliability and reduce delays.

Synchronisation of vehicles, docks, and equipment for unloading and loading goods will be required.

New organisations will need to be established for planning and managing the transfer facilities or hubs such as micro-consolidation centres and parcel locker banks associated with open multi-modal networks. These organisations will have a broader role than that of traditional logistics service providers.

Open and shared hubs are new types of facilities that require innovative organisations to operate and manage them effectively. This is due to the complexity of coordinating a number of independent shippers and carriers and organising loading and unloading docks as well as temporary storage areas.

Organising the operation of HCL hubs will involve the management of the resources available at these facilities and controlling the costs and revenue streams that are similar to those associated with operating urban consolidation centres.

Procedures for dynamically allocating vehicles to docks considering the availability of docks, load transfer equipment and the location of vehicles receiving loads will need to be developed. Improved systems for predicting the expected time of arrival (ETA) of vehicles at hubs will be also required. IoT will provide a means of monitoring the location of vehicles and their consignments as well as the status of loading docks and equipment.

There are a number of initiatives such as on-line platforms and network optimisation software that will need to be developed to facilitate HCL. On-line platforms will be required to effectively plan, design and operate HCL initiatives. On-line platforms are required to connect shippers and carriers as well as carriers and carriers. This allows the exchange of goods to be arranged and pricing to be negotiated as well as the level of service agreements to be made. Without on-line platforms, shippers and carriers cannot exchange information to allow effective sharing of transport and logistics resources. On-line auctions and systems for allocating freight vehicles to loads will provide an effective means of formalising the transfer of consignments between carriers.

Specialised software procedures for coordinating activities at transfer points to ensure high levels of reliability and efficiency will be required to implement HCL. Advanced loading dock booking and allocation systems will allow a high degree of coordination between carriers and hub operators. On-line loading dock booking systems are an example of the type of systems that will be required to effectively manage hubs to avoid delays to carriers.

Currently urban delivery routes are typically undertaken using a single mode of transport. This leads to low levels of consolidation and network efficiency. To reduce distances travelled and emissions by freight vehicles in urban areas, new optimisation procedures for designing collaborative networks based on multiple modes will need to be developed. Advanced software systems will be necessary to determine which hubs or transfer facilities should be used as well as what new modes (including electric vehicles) are available to minimise emissions and financial costs, and satisfy service delivery requirements.

12.3 EMERGING TECHNOLOGIES

12.3.1 Connected and autonomous vehicles

Autonomous vehicles described in Chapter 8 will lead to improvements in safety and efficiency. The development of driver assistance and automatic braking systems will be particularly important in turning situations to avoid crashes. Alert systems should consider driver reaction times based on vehicle trajectories based on sensors. Automatic braking systems are required when potential collisions are predicted. Collisions between turning trucks and cyclists generally have serious consequences for vulnerable road users. Due to driver error and blind spots, there is a need to adapt technologies to alert drivers to the presence of pedestrians and cyclists in the vicinity of trucks.

Autonomous vehicles provide substantial opportunities for reducing labour costs and emissions for last-kilometre deliveries. It is important to limit interactions with persons as well as challenging infrastructure. There is a need to develop procedures for determining routes for delivery robots to minimise conflicts with pedestrians and avoid difficult terrain. This will require integration of optimisation methods with advanced mapping systems and pedestrian monitoring systems.

Real-time communication between vehicles and infrastructure will lead to many benefits. Exchange of data between trucks and traffic signals allows more energy-efficient movement of trucks through signalised intersections. Procedures for determining optimal speeds will need to be developed to allow trucks to move more smoothly through signalised intersections considering the signal timing plans and traffic conditions. Future signal coordination systems could be designed to give priority to trucks that have high load factors or have engines that produce low or zero emissions to provide incentives for carriers.

Connected freight vehicles will also allow dynamic road pricing based on congestion levels, vehicle load factors and emissions. Advanced route guidance systems to allocate vans and trucks to loading zones and docks will be required.

Testbeds for developing and trialling connected vehicle technologies such as the Australian Integrated Multi-modal EcoSystem (AIMES) allow effective partnerships between industry and government to be established. Improved procedures for predicting travel times based on historical and real-time information such as congestion levels and incidents can be used to improve ETAs. Warning systems designed to alert drivers and respond to pedestrians and cyclists in the vicinity of vehicles have good potential for improving safety.

12.3.2 Electric vehicles

There is a need to incorporate vehicle charging within fleet management and operations for electric freight vehicles since this affects the financial and operational performance of carriers. Whilst downtime charging is common, other strategies used by carriers that involve charging during vehicles operating cycle including opportunity charging as well as intrusive strategies where routes are disrupted and trigger-based emergency charging. Improved decision support systems for determining the best charging strategies and operational decisions for fleet managers and drivers will be required.

The location and capacity of charging stations as well as the type of charging services offered will be important decisions for energy providers in urban areas. Behavioural models will need to be developed for better understanding and predicting the adoption of charging strategies used by carriers.

Dynamic wireless charging provides an opportunity to recharge vehicles as they travel without the need to stop that could address issues associated with the lack of charging infrastructure as well as even out the demand on electricity grids and allow solar energy to be utilised without the need for storage.

12.3.3 Digital twin (DT)

The digital-twin (DT) concept is an emerging technology that aims to provide a virtual system for replicating physical systems and their processes that allows the dynamic status of the environment and physical elements in the real world to be updated using various sources of information. Models are used to simulate scenarios and predict key indicators using data from regular updates.

Predicted changes in operating environments such as weather conditions and congestion levels can be incorporated into decision support. The status of assets in the real world such as the location of vehicles can be updated within models. Sensing and communication technologies allow real-time interaction and decision making with the latest state of the physical system providing feedback as part of the improvement cycle. Discrete event and agent-based simulation models can be combined with live data.

A large amount of data is required to create a digital twin since they are cyber copy of well-defined, real-world systems. Physical twins of urban freight networks include transport networks (including links and nodes), demand (supply and service points) as well as containers (load units), hubs and movers (vehicles). Geographic Information Systems (GIS) provide procedures for managing data as well as visualisation and animation. Databases are updated in real time with simulation and optimisation models connected to sensors. Digital twins require sharing of real-time information that will need a high level of data security and trust. Blockchain and smart contracts may provide an opportunity to overcome these challenges.

In Japan, the Ministry of Land, Infrastructure, Transport and Tourism has already created a data platform that allows 3D topographical maps to be linked with 3D city models (PLATEAU). In the future digital twins are planned to be created for 56 cities to promote business efficiency and smart cities, through industry-academia-government collaboration.

12.3.4 Data

A large amount of urban logistics involves goods being stored, sorted and distributed. Between shippers and receivers, goods are typically carried by a number of transport organisations and are transferred at several hubs. This creates the need to exchange data between companies involved during the first mile, sortation, and last mile. New data standards such as GS1's Scan4Transport (Scan4Transport - GS1 Australia) will provide a practical means of capturing data as goods move along the supply chain, simplifying integration as well as improving freight visibility and agility. Efficient exchange of data between organisations will be vital for realising hyperconnected city logistics.

The growth of near real-time data collected from sensor networks provides an opportunity for new data mining approaches to create smarter cities. Data mining methods provide opportunities for information to be gleaned from vast amounts of data. In contrast to previous scientific approaches, data mining allows new theories to be developed based on real-world observations.

12.4 INTEGRATED PLATFORM

12.4.1 Platform characteristics

To address the growing challenges associated with productivity, sustainability, and resilience, future analytics systems will need to be based on more integrated platforms. Such platforms will need to consider multiple stakeholders or agents, including emerging organisations such as PI companies and entities the crowd as well as traditional stakeholders such as shippers, receivers, carriers, administrators and residents. A range of performance criteria for private sector such as viability and practicality as well as for administrators such as environmental and social impacts will need to be predicted.

Integrated platforms will allow a range of new HCL facilities and services such as micro-consolidation centres (MCC), joint delivery systems (JDS), crowdshipping, and parcel lockers to be designed and evaluated for implementation in urban areas. Such platforms will provide users and policymakers of HCL systems with decision-making support in operation, planning, regulations, and investments and help them understand how their behaviour, in turn, affects the system. An integrated set of modelling approaches will be required to evaluate and recommend the best type of PI initiatives and (public) policies for deriving efficient and sustainable urban freight networks. The modelling platform will need to consist of stated preference methods (SPMs), gamification, discrete choice models (DCMs), learning models, agent-based simulation models (ABSMs), optimisation models, and machine learning (ML) algorithms.

To evaluate policies for promoting HCL, administrators will need tools for predicting the price of services offered by HCL platforms and understanding how HCL organisations such as hub operators engage with carriers (including the crowd). The learning behaviour of stakeholders considering experience and knowledge will also need to be incorporated.

Future modelling platforms will need to combine a range of modelling approaches including behavioural choice, simulation and optimisation. Technologies such as digital twins will provide data integration and visualisation.

There is also a need to predict and appraise the benefits of implementing city logistics initiatives in both short and long terms, relative to the level of engagement of different stakeholders. Consideration of the timing of decisions (operational, tactical, and strategic) for stakeholders will need to be included.

Agent-based simulation provides an overarching modelling approach for representing the dynamic interactions between stakeholders. Behavioural choice models can be used to predict the decisions of stakeholders. Optimisation models will be required to predict how resources can be managed by key stakeholders. Serious games will also be an integral element in future analytics platforms.

Decisions made by the agents can be categorised into operational, tactical, and strategic planning based on their decision timelines. For example, the daily operational planning of a carrier includes increasing/decreasing/keeping the price (to sell or buy delivery service), delivering goods by itself or using JDS with MCC, and choosing the delivery modes (*e.g.*, van, bike, electrical vehicle, drones, on foot) and routes. To make a decision (and take an action), the carrier needs to interpret the state of the system, which includes the information about the resources and prices offered by the JDS and the parcel locker company, the policy enforced by the administrator, the orders received from shippers as well as traffic conditions.

Locker companies will need to decide the location and capacity to install parcel lockers as a strategic plan, which may occur at the beginning of a year or quarter. Another example is the administrator may develop strategic plans on restricting or charging couriers for travelling in certain areas such as central business districts and providing subsidies to the development of HCL platforms in order to encourage consolidation and joint deliveries, which will reduce delivery trips, decrease environmental impacts, and improve distribution efficiency and sustainability. ABSM will assume shippers choose services based on prices and delivery preferences and times.

Future modelling platforms will be required to evaluate the scenarios which represent traditional stakeholders and HCL organisations operating at different levels of interconnectivity. The comparisons of the scenarios will provide comprehensive assessments of the benefits derived from incorporating the HCL solutions and sharing of information, service, and resources on the efficiency of freight movements, the environmental footprint, the need for motorisation of vehicles, the number of empty/low-load vehicles in city areas, congestion, fast and reliable response delivery to stochastic demand, etc. ABS will be used as an environmental model in machining learning (e.g., neural networks, reinforcement learning) and simulation-based optimisation to identify the critical success factors relating to the configuration and coordination of PI organisations.

Attention will need to be placed on the assessment of the central role of administrators and regulators (governments) as the engagement of resourceful governments can greatly facilitate the conceptualisation of HCL to tackle logistics problems. A range of scenarios for the involvement of the administrator will need to be considered.

Under different policy scenarios, modelling platforms will need to consider two cases in terms of how other agents adapt to policies: (i) when the inputs from the choice and optimisation models can fully characterise the decisions of an agent; (ii) when the agents are enabled to conduct further learning in the ABSM. For case (ii), reinforcement learning will need to be integrated into agents' decision making.

Urban freight systems in the future will involve the participation of many entities such as transport service providers, parcel locker companies, retailers, crowdsourcing platforms, and administrators, with each actor generating millions of data points daily. Effective operations and management of urban freight systems rely on reliable and timely data. Particularly, the administration, coordination, and optimisation of resources will benefit from advanced analytics and computationally efficient algorithms that incorporate ML and artificial intelligence (AI) techniques. ML and AI methods combined with IoT and digital-twin technologies have the potential to (a) facilitate intelligent and autonomous decision making, (b) learn from and predict the behaviour of the system and its components (*e.g.*, trajectory modelling), (c) design and manage logistics, in particular HCL, facilities, and (d) enable real-time and seamless communication across the platform. Blockchain can further expedite great trust among stakeholders and customers, supporting new services and business models.

Transforming existing urban freight networks to HCL systems will require the participation and efforts of all stakeholders, traditional and new actors and regulators, and require a long timeframe. Future analytics platforms will require the prediction of the responses to new PI-CL solutions for all major stakeholders and the assessment of impacts on the urban freight network such as sustainability, as well as the overall transport network performance such as urban congestion. Advances in gamification, ABSM, ML and data-driven optimisation will need to be utilised.

Gamification, based on serious gaming principles, can be adapted to collect data on decision making by shippers, carriers, and new HCL organisations, which can overcome the obstacle of lack of open data. Stated preference methods can be used to design realistic, competitive, and interactive settings for developing choice models. A set of data-driven optimisation models can also be developed to represent the decisions of the stakeholders at the different planning levels. In terms of optimisation development, as described in Chapter 5, ABS provide a framework for the advance of knowledge in the interactions between ML and optimisation, as the results of each individual decision-making problem will be analysed within the broad freight transport context represented. This is particularly true for tactical and strategic problems where data-driven stochastic models will interact over time within the ABS, allowing the possibility of studying the effect of recourse actions in a controlled environment and providing the possibility to evaluate new ideas in algorithmic development.

12.4.2 Agent-based simulation models (ABSM)

Agent-based simulation models (ABSM) provide a means to integrate advanced modelling methods. This will allow financial viability and flexibility to be investigated by representing the interactions between stakeholders' decisions over time, with variable demand levels, reactions, incorporating traditional stakeholders and HCL-orientated agents. Learning agents, in particular ML/AI-based ones, allow stakeholders to be more adaptive and responsive in their decision making with re-optimisation over extended periods. Learning effects can be incorporated via simulated feedback on the performance of previous decisions and the behaviour of all stakeholders. In addition to the prediction of benefits that HCL solutions can yield over time, which can motivate stakeholders' participation and engagement, expected outcomes include the identification of the role of each major stakeholder can play in smooth and fast transformation to HCL systems. For actors in the private sector, methods could be obtained for determining how best decisions on such as capacity, pricing, service sharing, can be used to improve their performance (e.g., profitability, resilience). For regulators, knowledge will be advanced about the design and implementation of freight policies and regulations to support the development of HCL.

Agent-based simulations of HCL solutions will require simplified but realistic representations of decisions of the key agents involved in the urban freight system. This will need high-quality behavioural data from a diverse set of stakeholders in the system which is not currently available. Agents within the ABSM have the capability to learn and adapt their behavioural rules based on their own experience and goals as well as their interactions with the system and other entities.

HCL organisations can be represented as new types of agents and their is a need to study their behaviour and interactions. Compared to the traditional stakeholders in urban logistics, HCL components can make more use of the capability of autonomous agents as they require sharing more information, resources, and business, have more interactions and complex and complicated decisions to make. The ABSM can take the outcomes of choice and optimisation models as inputs and provide a platform, which integrates these inputs with ML, to study the interactions and synergies among various stakeholders and the emergent behaviour, which is hard to predict from the agents' individual behavioural rules. Moreover, it will enable the study of the role of administrators as HCL facilitators through various logistics policy measures during the evolution of the conventional urban logistics network towards a hyperconnected, shared system.

12.4.3 Gamification

There is a need to develop improved methods for understanding the behaviour of urban freight stakeholders in response to HCL solutions. Gamification or serious games is an emerging research method for data acquisition in urban freight and logistics. It is becoming more recognised as an innovative data collection tool for building choice and simulation models.

Interactive games allow data to be collected passively to understand how stakeholders will respond and utilise new initiatives. They can be used to collect data to characterise the actions of traditional and new major stakeholders for developing choice and optimisation models to represent their decision-making processes and to support the design, evaluation, and implementation of HCL solutions.

Games provide a practical and cost-effective way of understanding urban freight systems allowing choices to be observed in a simplified representation of complex and dynamic systems. Using games, choice situations can be constructed to reflect the types of new services that HCL initiatives will offer, and stakeholders can make decisions about managing their limited resources and operations with predefined rules that reflect the nature of engagement with the new services providing a statistically rigorous method for predicting the behaviour of each player.

Serious games have great potential for observing, logging, analysing, and modelling the behaviour of city logistics stakeholders. The data collected from games will allow discrete choice models for both logistics demand and supply to be constructed that will be used to replicate the behaviour in ABSM.

A common problem is the lack of sufficient data for planning and optimisation of urban freight systems which is a major obstacle towards laying the groundwork for the HCL system. Stated preference methods have recently been used to examine the attitudes and preferences of carriers and receivers towards consolidated delivery from UCCs in inner-city areas. However, DCMs developed based on data collected from the games will gain insights and explain the choices related to how shippers and carriers interact with new HCL organisations, utilise their services, and respond to related policies, which will enable the ABSM to capture the system dynamics before and after the conceptualisation of HCL.

Games can be used to investigate sets of decisions to be deployed in the ABSM such as:

i. how carriers and receivers will utilise HCL facilities such as parcel lockers under various types of pricing schemes as well as specific locations (Pan et al., 2021; Guo et al., 2021),
ii. how shippers and carriers will engage with Delivery as a Service (DaaS) organisations with respect to offering and accepting jobs under different pricing, service level requirements, incentives, and freight policies, and
iii. how the public will engage with crowdshipping activities and hosting parcel lockers.

The data collected in game experiments can be used to identify and quantify the key attribute components that influence the usage of HCL solutions. Once gamification experiments are conducted and data is collected and analysed, DCMs can be developed based on observed choice decisions made by players.

12.4.4 Optimisation

Optimisation models described in Chapter 4 have been widely studied and used for informing the decision making and methodology of major actors in city logistics. However, optimisation models cannot directly or easily to used represent the decision making of the actors in the HCL system, because the presence of HCL solutions introduces new decisions and opportunities for the actors and increases the complexity of their problems.

Common decisions made by carriers, HCL organisations, and administrators can be classified into strategic, tactical, and operational levels. Consider the operational planning for a carrier for example. In current urban logistics networks, the carrier's primary task involves the delivery route plan and the transport mode choice (among traditional vans, trucks, and environmental-friendly electric vehicles, bikes, *etc.*). In the HCL system, the carrier needs also to consider when and at what price to sell its delivery service if it has spare vehicles or space in vehicles and to buy service if it operates at or over capacity, and choose business partners, if multiple buyers or sellers are available (*e.g.*, crowdshipping and UCC companies). Therefore, new business models and optimisation approaches are required.

For each agent, decision problems, which integrate the current operations and new business processes, can be modelled and solved using the most appropriate optimisation technique. This suite of models will be required to assess and analyse each decision-making process from the perspective of an individual agent, validating the interactions of this agent when it is incorporated into the complete simulation model. The goal is not to model the whole network (as this is the role of the ABSM) but represent the core of each decision-making situation, focusing on what is needed for the simulation.

In the past, Mixed-Integer Linear Programming (MILP) and metaheuristics described in Chapter 4 have assumed full knowledge using deterministic data, and their solutions can provide fast and useful insights into what transformative changes are essential for establishing horizontal collaboration with other logistics organisations considering internal and external factors. However, in the future they will need to be extended to incorporate uncertainties by means of possible future scenarios. The optimisation goal then becomes to minimise an expected cost value (stochastic programming, chance-constrained programming) or to maximise performance under worst-case situations (robust optimisation). Solutions will need to inform stakeholders' planning and operations in a complex planning environment. Obtaining data for solving these optimisation models in this context is an arduous task in itself. Data-driven optimisation techniques, where machine learning approaches are used to format the data into the needed scope, have good potential. Optimisation models can be used not only to validate and improve heuristic decisions taken by stakeholders but also to incorporate uncertain information scoped from serious games and other sources.

An important modelling aspect is the choice of the appropriate scope for each decision problem such as decisions on the number, location, sizes, and functionality of parcel locker facilities to be used as transfer points in last-mile distribution with joint delivery service and crowdshipping as well as collection points for customers. Existing works have studied similar problems where parcel lockers play a solitary role in JDS, crowdshipping, and collection services. The management of multi-functional parcel lockers that are open to all logistics players and users and explore whether such tactical/strategic decisions would be better or not synthesised to operational ones for improved levels of efficiency and robustness will need to be investigated.

12.4.5 Machine learning (ML)

Machine learning methods will be required to scope large amounts of data collected from gamification experiments and DCMs into the necessary format for optimisation methods. Integrating ML and optimisation algorithms is known as data-driven optimisation. Choice and optimisation models will need to be reflected in the agents' (stakeholders') decision-making strategies in the ABSM. ABSM can equip the dynamic agents with intelligent learning abilities (*e.g.*, dynamic ML algorithms) to complement their decision-making processes. Furthermore, the outcome of the ABSM can inform and cross-validate the gamification experiments and the optimisation models, creating a feedback loop between models. ABSM can be used as a testbed to examine administrators', as one agent type, intervention to steer the transformation from the existing network to the HCL system, as well as other agents' responses and emerging behaviour to city logistics initiatives and policies. Such a platform allows examination of the effectiveness of HCL facilities, services, protocols, and policies on short- and long-term economic, social, and environmental performance.

12.5 CONCLUSIONS

To improve the sustainability and efficiency of urban logistics systems a more integrated approach to urban logistics network planning and operation will be necessary. Hyperconnected City Logistics (HCL) provides a framework for creating more open and shared urban logistics systems that can increase consolidation levels and make a range of low emission vehicles more practical and viable.

However, to facilitate the implementation HCL a range of new models will be required based on real-time data. Machine learning has good potential for improving decision making by incorporating historical as well as real-time data. To effectively evaluate public sector policies an integrated modelling platform will be necessary where agent-based simulation models can be developed that incorporate behavioural choice models as well as optimisation models.

REFERENCES

ALICE (2015). Urban Freight, *Research and Innovation Roadmap*, Alliance for Logistics Innovation through Collaboration in Europe, European Road Transport Research Advisory Council (ERTRAC), Brussels.

Crainic, T.G. and Montreuil, B. (2016). Physical internet enabled hyperconnected city logistics, *Transportation Research Procedia*, 12, 383–398.

Guo, C., Thompson, R.G., Foliente, G. and Kong, X.T.R. (2021). An auction-enabled collaborative routing mechanism for omnichannel on-demand logistics through transshipment, *Transportation Research Part E*, **146**, 102206.

Montreuil, B. (2011). Toward a physical internet: Meeting the global logistics sustainability grand challenge, *Logistics Research*, 3, 71–87.

Pan, S., Zhang, L., Thompson, R.G. and Ghaderi, H. (2021). A parcel network flow approach for joint delivery networks using parcel lockers, *International Journal of Production Research*, 59, 2090–2115.

Taniguchi, E. and Thompson, R.G. (2015). *City Logistics: Mapping the Future*, In: Taniguchi, E. and Thompson, R.G., (eds.), CRC Press, Taylor & Francis.

Index

Note: Page numbers in **bold** refer to tables and those in *italics* to figures.

Printed in the United States
by Baker & Taylor Publisher Services

Printed in the United States
by Baker & Taylor Publisher Services